FLOWERING PLANTS
OF THE
SOUTHWESTERN
WOODLANDS

TERALENE S. FOXX & DOROTHY HOARD

Illustrations by Dorothy Hoard

Photographs by Teralene Foxx

1995

Otowi Crossing Press
Los Alamos, New Mexico

Copyright 1984 by Teralene S. Foxx and Dorothy Hoard
under title Flowers of the Southwestern Forests and Woodlands

All rights reserved, including the
right to reproduce this book
or portion thereof in any form.

Published by Otowi Crossing Press
1350 Central
Los Alamos, NM 87544

Library of Congress Cataloging in Publication Data

    Foxx, Teralene S.
    Flowering plants of the southwestern woodlands

    Bibliography: p.   Includes Index
    1. Forest flora—Southwest—Identification.
    2. Botany—Southwest. 3. Plants—Identification.
    I. Hoard, Dorothy.   II. Title.
    Library of Congress Catalogue Card Number: 95-069166

ISBN 0-9645703-1-9

Cover illustrations by Dorothy Hoard

Second Printing 1995

Preface

In 1984, we first published the contents of this book under the title *Flowers of the Southwestern Forests and Woodlands.* We had spent 10 years doing research for the book in the Jemez Mountains and on the Pajarito Plateau, a table-like extension of the eastern flank of the Jemez Mountains in Northern New Mexico. During that time we made collections in Bandelier National Monument, Los Alamos National Laboratory, and Los Alamos County. Our intention when developing this book was to help the novice understand the wonderful world of wild plants in our area. We soon found, however, that it could be used in mountain ranges throughout New Mexico and Arizona. Its usefulness has expanded beyond our original vision.

The scope of this book is the wooded areas in the mountains of the Southwest: pinyon-juniper woodland and the ponderosa pine, mixed conifer, and spruce-fir forests. It also includes plants that are found in streamside and meadow habitats dispersed throughout these forest types. The book includes the majority of plant families and most of the common plant species of the Pajarito Plateau and the Jemez Mountains and similar mountain ranges in Arizona and New Mexico.

There are over 900 species on the plateau; this book covers about 450 species. As with any book that is not comprehensive, there are problems of omission, but we believe we present a representative sample. As authority for scientific names and classification, we have used *A Flora of New Mexico* by Martin and Hutchins (Germany, J. Cramer, 1981). This is the most recent and comprehensive study of flora of the state. For additional information on plant species of the plateau, see *A Checklist of Vascular Plants of the Pajarito Plateau and Jemez Mountains* by Foxx and Tierney (New Mexico, Los Alamos National Laboratory, 1984).

This book was written for the beginner, the person who wants to do a systematic study of plants beyond the simple picture book but does not have the technical knowledge to use botanical floras. Since my undergraduate days at the College of Idaho, I have wanted to introduce the novice to the world of plants. In 1970, I devised a course for continuing education at the College of Santa Fe and taught it for several years, both in Santa Fe and at the Los Alamos branch of the University of New Mexico. One of my earliest and most enthusiastic students was Dorothy Hoard. We have pooled our efforts and have enjoyed a close collaboration ever since.

One of the most difficult tasks in teaching a plant identification course was finding a suitable text. Most comprehensive books were far too complicated and technical; beginning students soon got discouraged. I began developing simple keys to use with my classes. These keys have evolved into this book. Since it was first published, I have taught several hundred students. Many of these students have used the book as a stepping stone to more advanced texts on plant identification. Others have continued the joy of discovery using the keys we developed. Beyond its use as a text, it has been a baseline for studies we have done at Los Alamos National Laboratory and Bandelier National Monument.

A book is not just the contribution of one or two people. It takes many people to make a book successful. First and foremost are personnel of the institutions that gave me the opportunity to teach courses in plant identification: College of Santa Fe, Los Alamos branch of the University of New Mexico, and Ghost Ranch. I also appreciate the opportunities for research provided to me by Los Alamos National Laboratory and Bandelier National Monument. Over the years my many colleagues have challenged and encouraged me with their ideas and questions. Of special note are John Lissoway, Craig Allen, and John D. Hunter of Bandelier National Monument and the biological team at Los Alamos National Laboratory.

I owe particular thanks to the many students I have used as guinea pigs in my attempt to make this book useful to the novice. Dorothy Hoard and I thank the following people who read and criticized all or portions of the original text: Timothy Fischer, Donald Hoard, William Martin, Richard Peters, Relf Price, and Gail Tierney. We thank Ken Ewing, who proofread the manuscript. Becky Shankland, our editor, spent many hours reworking the manuscript to make it readable, consistent, and grammatical. We thank our families—our children, now adults, who spent many hours waiting on the trails while we looked at wild flowers, and our husbands, Jim Foxx and Donald Hoard, who stoically endured this labor of love.

Our special thanks go to Colleen Olinger of Otowi Crossing Press for reprinting this book as part of her effort to provide information to residents and visitors alike on the natural history of the Southwest we all cherish.

Teralene Foxx
Los Alamos, New Mexico
May 1995

CONTENTS

PREFACE iii

INTRODUCTION 1

HOW TO USE THIS BOOK 4

PLANT STRUCTURE 5

HINTS FOR PLANT IDENTIFICATION 6

POISONOUS PLANTS 9

DETERMINING WHAT CHAPTER TO USE 10

KEYS
- Vines and Trailing Plants 12
- Trees 17
- Shrubs 28
- Parasites and Saprophytes 48
- Horsetails 51
- Grasses 52
- Cattails, Rushes, and Sedges 65
- Cacti 68
- Ferns 72
- Composites 74
- Showy Monocots 99
- Weeds 107
- Herbaceous (non-woody) Dicots 112

GLOSSARY 188

SELECTED REFERENCES 195

PRONUNCIATION GUIDE 196

INDEX 200

To see a World in a grain of sand
    And a Heaven in a wild flower
Hold Infinity in the palm of your hand
    And Eternity in an hour.

                      William Blake

INTRODUCTION

PLANT IDENTIFICATION

Plant identification can be fun. It does take practice and it does take time. You will need a minimum of equipment, but several items are essential for this wonderful hobby. These are:

    a good reference book

    a hand lens or magnifying glass (10X to 20X)

    a ruler with both English and metric units

    patience and persistence.

PLANT CLASSIFICATION

The science of classification is called taxonomy. The system of taxonomy currently in use in biology was developed by Swedish botanist Linnaeus in the 1700s. Living organisms are grouped according to their reproductive relationships. The groups are ranked as follows:

    species (lowest rank): plants or animals so like one another they can interbreed

    genus: a group of related species

    family: a group of related genera

    order: a group of related families

    class: a group of related orders

    phylum: a group of related classes

    category (highest rank): a group of related phyla.

In some cases groups are subdivided, e.g., the orders belonging to a class may be divided among two or more subclasses.

When identifying plants, however, it is rarely necessary to pay much attention to such distinctions. This book stresses the three lowest groups—species, genus, family and two subclasses. Once one recognizes the characteristics of a family, it becomes easier to identify member species of that family.

This book uses both common and scientific names. Scientific names have the advantage that only one plant species has that particular name. Common names may refer to several plants, often in entirely different families. Use of the scientific name allows for more precise communication.

The scientific designation of a plant always includes two words: the genus name and the species name. These names are usually either italicized or underlined because they are Latinized. As an example, corn is known scientifically as *Zea mays*.

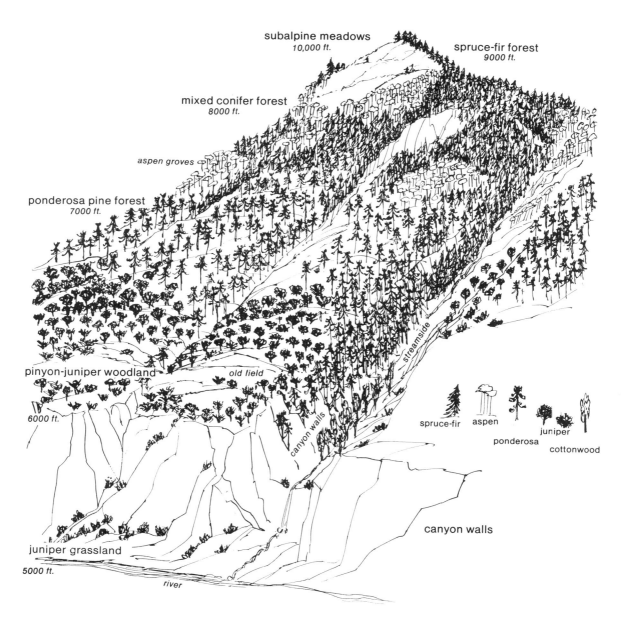

PLANT COMMUNITIES AND HABITATS

PLANT COMMUNITIES

In ecology, a community refers to all the plants and animals in an area. The term is usually applied to organisms needing similar conditions for life. In the case of plants these conditions may be specific soils and or/moisture and temperature requirements. Large stands of a specific plant species will often be found at certain elevations or on similar soil types. A cluster of one species of plant becomes known as a plant community. The name of the plant community is dependent on the overstory or dominant tree species. If trees are not the dominant overstory, the plant community is named after the dominant shrub or grass. Areas devoid of trees or shrubs may be called grasslands or meadows. On the Pajarito Plateau and throughout the high elevations of the Southwest, six plant communities are defined. They are juniper grasslands, pinyon-juniper woodlands, ponderosa pine forests, mixed conifer forests, spruce-fir forests, and subalpine meadows. Within these communities, factors such as moisture may affect species composition. Locations within the plant community that vary because of environmental conditions are denoted as habitats (i. e. old field, aspen groves, streamside). Streamside (riparian) habitats are dispersed throughout the plant communities. Because of moisture, very different plant species survive along streams than in the open forest.

The pinyon-juniper woodland ranges from elevations of 6000 feet to approximately 7000 feet, and is dominated by pinyon and juniper. This woodland is somewhat arid and trees are widely spaced. Common grass species present include blue grama, galleta, and little bluestem.

The ponderosa pine forest begins at elevations of approximately 7000 feet. Between 7000 and 8000 feet these pines form almost pure stands. At lower elevations they are found on north-facing canyon walls and along streams. At higher elevations ponderosa pines are scattered through the mixed conifer forest. Dominant grass species here include mountain muhly, pine dropseed, and little bluestem.

The mixed conifer forest ranges from the 8000- to 9000-foot elevations. Interspersed in this high elevation forest are Douglas fir, white fir, corkbark fir, limber pine, ponderosa pine and groves of aspen. Much of the area has an understory of bearberry, creeping barberry, and various grasses and forbs.

The spruce-fir forest occurs in small stands near the summits of the mountains. Engelmann spruce, white fir, and corkbark fir predominate. These trees often form dense stands with little understory.

Subalpine meadows are found at the tops of the higher peaks at about 10,000 feet, generally on south-facing slopes. A few trees--Douglas fir, aspen, and spruce--encroach on the meadows, and are misshapen by the winds. The meadows are dominated by oatgrass and brome grass.

Riparian areas of the plateau provide habitats for various water-loving species: alder, birch, willow, and poplar. In canyons, near perennial streams, many species of trees, shrubs, grasses, and wildflowers occur. A more drought-resistant flora grows at the lower elevations where streams are intermittent and the stream banks become more arid.

HOW TO USE THIS BOOK

This book uses a systematic method of plant identification called dichotomous keys. Such keys are made up of couplets--pairs of contrasting statements, each statement bearing the same number. Beneath each statement of a couplet is usually another couplet of statements, each of which has a higher number; beneath each statement of the second couplet may be another couplet, and so on. Ultimately, though, the search will end with a short paragraph describing in some detail the characteristics of a single plant species. These paragraphs are supplemented with line drawings of the species described. Compare the features mentioned in both statements of a couplet with those of the plant you are trying to identify, choose the statement which better describes the plant, and proceed to the couplet following the statement chosen. If each choice has been made correctly, you should arrive at the proper plant description.

Many plants are not easy to identify. It is necessary to read each complete statement and carefully study the plant. If at any time neither statement of a couplet seems to describe the plant, return to the beginning and try again, choosing a different branch of a couplet at the confusing places. If any statement has unfamiliar terms, consult the glossary.

You must proceed systematically by following these steps:

STEP 1: Turn to DETERMINING WHICH CHAPTER TO USE, p. 10.

STEP 2: Study the plant. If the specimen is a flower, did it come from a shrub? Is the flower that of a monocot, a dicot, a composite? (Consult the glossary if these terms are unfamiliar.)

STEP 3: Turn to what seems to be the proper chapter. Note that some chapters are divided into sections. Go directly to the section which best describes the plant.

STEP 4: Examine BOTH statements of the couplet numbered 1. (This will often involve searching through several pages for the second statement.) Choose the statement which better describes the plant. Proceed in the same way with the couplet beneath the statement you have chosen.

STEP 5: After reaching a final description (no further couplets), compare the plant with the drawing. If you see minor differences, you may have correctly identified the genus of your plant, but its species may be different from the one described. If there are major discrepancies between the written description and your specimen, and the drawing does not resemble it, you must go back and try again.

STEP 6: As you become familiar with the characteristics of the various groups, you will be able to shortcut the initial steps of the keys.

In a few places you will find that a choice leads to rather sketchy descriptions of several species, rather than a single one. This occurs in two instances. Sometimes hybridization (crossing) between species within a genus is frequent. (In such cases, you should be satisfied with having identified the genus of the plant--experts are not always sure about the species, either.) You will also find brief descriptions of several species when many similar members of a genus occur together on the Plateau. It just wasn't practical to discuss all the species (and the somewhat obscure features separating them) in a book of this size.

As will quickly be evident, both English and metric measurements are used in this book. Any measurements less than 1 ft (30 cm) are in centimeters; measurements greater than 1 ft (30 cm) are in feet. All measurements represent length unless otherwise specified.

PLANT STRUCTURE

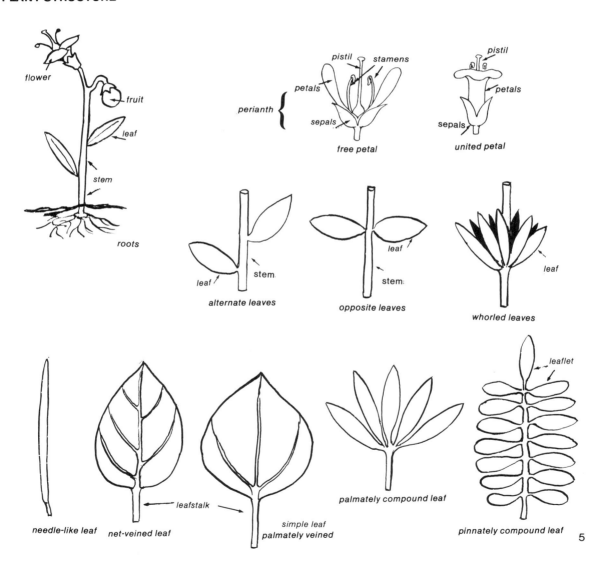

HINTS FOR PLANT IDENTIFICATION

Keying plants can be time consuming and frustrating. It should be the goal of every serious student of plant identification to reduce the keying time to a minimum by learning the attributes of families or genera. Taxonomy is based on specific morphological characteristics; therefore, knowledge of those characteristics can be used to go directly to the family rather than spending time searching through the key. Some families or genera have unique characteristics useful in recognizing the plant group. For example, only three families in this book have square stems: the mint, the madder, and the vervain families. Knowing these characteristics eliminates many other families. Also, knowing other family attributes can further aid or narrow the identification. The mint family has asymmetrical flowers, the madder has symmetrical flowers, and the vervain family has slightly asymmetrical flowers.

The hints given below are only aids, not guarantees. There may be exceptions, and often are. But knowledge of the specific characteristics can help in identification and make it more fun.

Flower symmetry.

Symmetry refers to the position of sepals and petals. Plants having all sepals and petals similar in arrangement, shape, and size are *symmetrical*. Such flowers can be divided into equal halves by a vertical plane at various places along the face of the flower. Plants having petals dissimilar in size and shape are *asymmetrical*. Such flowers may be two-lipped or may have one petal spurred. Below are listed families found in this book having asymmetrical flowers.

Orchid family ORCHIDACEAE
Mint family LABIATAE
Pea family LEGUMINOSAE
Figwort family SCROPHULARIACEAE
Violet family VIOLACEAE
Vervain family VERBENACEAE
Buttercup family RANUNCULACEAE (some members)

Flower completeness.

In some cases not all floral parts are present. Petals, stamens, or pistils may be missing. Flowers that do not have petals are said to be *apetalous*. In such cases the sepals may be showy and petal-like. Other flowers may be missing either male (stamens) or female (pistil) parts. If male and female flowers are on one plant, the plant is *monoecious*. If male and female flowers are on separate plants, the plant is *dioecious*. Below are listed families which are apetalous, monoecious, and dioecious.

Apetalous flowers

Four o'clock family NYCTAGINACEAE
Buttercup family RANUNCULACEAE (some members)
Buckwheat family POLYGONACEAE
Goosefoot family CHENOPODIACEAE

Monoecious flowers

Gourd family CUCURBITACEAE
Birch family BETULACEAE
Pine family PINACEAE

Dioecious flowers

Willow family SALICACEAE
Cypress family CUPRESSACEAE
Sunflower family COMPOSITAE (some members)

Floral formula.

The floral formula is a numerical description of the parts of a flower, starting from the sepals and counting inward to the pistil (female part of the flower). The number and position of the flower parts are two of the key taxonomic characteristics used in plant identification. Plants in the subclasses Monocotyledonae (monocots) have flower parts in threes or multiples of three (i.e. six). Flowers in the subclass Dicotyledonae (dicots) have flower parts in fours or fives or multiples of these numbers (i.e. eight, ten). Dicots often have more than ten stamens or pistils. Such plants are said to have *many* stamens or *many* pistils. Because the subclass Dicotyledonae is large, it is useful to learn some unique floral formulas. The most common floral arrangement in this subclass is five sepals, five petals, five to ten stamens, and one pistil. A few families, however, have floral formulas based upon four or other numbers. Below are some families and their floral formulas.

Floral formulas based on 4

Evening-primrose family ONAGRACEAE 4,4,8, 4
Mustard family CRUCIFERAE 4,4,6,4
Caper family CAPPARIDACEAE 4,4,6,4

Floral formulas of 5, 5, many, many

Rose family ROSACEAE
Buttercup family RANUNCULACEAE
Mallow family MALVACEAE

Floral formulas with 2 sepals

Fumitory family FUMARIACEAE

Plant odor and sap color.

Most plants have colorless cell sap. In others it may be milky or colored. The sap will exude from stems or leaves when they are severed or crushed. Some plants may produce distinctive odors when parts are crushed. Below are listed plants with milky sap or distinctive odors.

Sap milky

Dogbane family APOCYNACEAE
Milkweed family ASCLEPIADACEAE
Sunflower family COMPOSITAE (some members)

Characteristic odor

    Geranium family GERANIACEAE
        Mint family LABIATAE
        Rue family RUBIACEAE
Sunflower family COMPOSITAE (some members)
Phlox family POLEMONIACEAE (Polemonium spp.)

Stem shape.

The shape of the stem can be an important identifying feature. To determine the shape, roll the stem gently between your fingers. Most stems are round, but a few families characteristically have stems square or triangular in cross section. Others have stems with nodes. Below are listed families having these characteristics.

Square stems

Mint family LABIATAE
Madder family RUTACEAE
Vervain family VERBENACEAE

Triangular stems

Sedge family CYPERACEAE

Stems with nodes

Grass family GRAMINEAE

Fusion of petals.

The floral parts may be free or fused either at the base or into a variety of shapes. It is useful to learn the families having free petals and those having united petals. They are too numerous to list here.

Differences between subclasses.

It is important to distinquish first the difference between the two subclasses of plants, the monocots and dicots. Monocots have floral parts in threes or multiples of three, leaves with parallel veins, and fibrous roots. Dicots have floral parts in fours or fives or multiples of these numbers, leaves variously veined, and taproots. All three characteristics will be important in placing plants in the proper subclass. Occasionally dicots have leaves appearing parallel veined but other features will place them within the subclass.

POISONOUS PLANTS

It is important to become familiar with poisonous plants, particularly those known to cause a dermatitis, before beginning to identify plants. Most plants are harmless; others may cause discomfort or even death when eaten. Still others may cause an uncomfortable skin rash when touched. This plate illustrates some common poisonous plants.

This book does not advocate the eating of native plants. Information on plant lore is included only for interest. Those curious about poisonous and edible plants should find Harrington's *Edible Native Plants of the Rocky Mountains* useful. Those desiring more information on herbal uses of plants might begin with Moore's *Medicinal Plants of the Mountain West*.

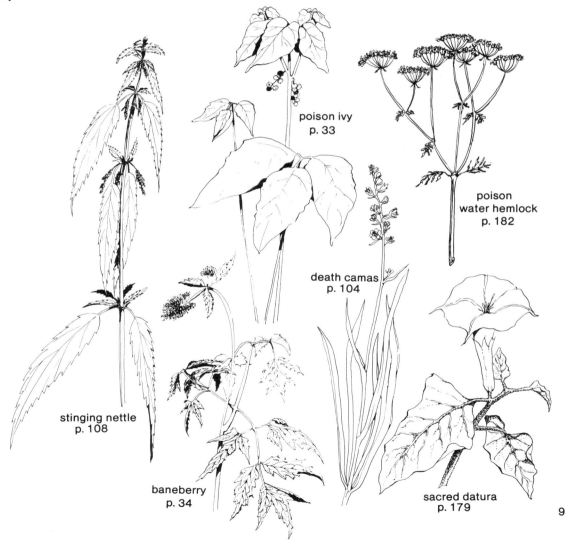

poison ivy p. 33

poison water hemlock p. 182

death camas p. 104

stinging nettle p. 108

baneberry p. 34

sacred datura p. 179

DETERMINING WHICH CHAPTER TO USE

If your plant is:

1. Woody plant.

    2. Plant vining, trailing, or twining.

                  Go to VINES AND TRAILING PLANTS, p. 12.

    2. Plant neither vine-like, trailing, nor twining.

        3. Plant with one trunk greater than 5 cm in diameter, mature individuals at least 10 ft tall.

                  Go to TREES, p. 17.

        3. Plant with multiple stems, each less than 5 cm in diameter, mature individuals less than 10 ft tall.

                  Go to SHRUBS, p. 28.

1. Herbaceous (non-woody) plant.

    4. Plant vine-like, trailing, or twining.

                  Go to VINES AND TRAILING PLANTS, p. 12.

4. Plant neither vine-like, trailing, nor twining.

    5. Plant without green color.

                  Go to PARASITES AND SAPROPHYTES, p. 48.

    5. Plant with green color.

        6. Plant with conspicuously jointed stems, or grass-like.

            7. Plant with conspicuously jointed stems.

                  Go to HORSETAILS, p. 51.

            7. Plant grass-like.

                8. Plant stems with swollen nodes.

                  Go to GRASSES, p. 52.

                8. Plant stems smooth.

                  Go to CATTAILS, RUSHES, AND SEDGES, p. 65.

        6. Plant neither grass-like nor with conspicuously jointed stems.

9. Plant with specialized stems bearing spines.

Go to CACTI, p. 68.

9. Plant without spiny, fleshy stems.

    10. Plant fern-like. Some leaves with raised dots on underside.

Go to FERNS, p. 72.

    10. Plant not fern-like. Leaves without raised dots on underside.

        11. Flowers in dense, compact, heads having ray and/or disk flowers. Flower heads resembling a sunflower, dandelion, or thistle.

Go to COMPOSITES, p. 74.

        11. Flowers not in dense compact heads.

            12. Plant with flower parts in 3 or multiples of 3. Leaves parallel veined.

Go to SHOWY MONOCOTS, p. 99

            12. Plant with flower parts in 4 or 5 or multiples of 4 or 5. Leaves net veined.

                13. Plant without showy flowers. Found in gardens, along trails and roadsides.

Go to WEEDS, p. 107

                13. Plant with showy flowers.

Go to HERBACEOUS DICOTS, p. 112

VINES AND TRAILING PLANTS

A *vine* can be defined as a plant which climbs or scrambles on some support. The stem does not stand upright by itself. A *trailing* plant lies flat on the ground. A *twining* plant grows upward by coiling around a support.

1. Stems woody or firm.

    2. Leaves opposite.

        3. Leaves simple, palmately lobed, sticky on underside. Fruit cone-like and papery, strong-smelling.

            Perennial with trailing or twining stems, rough-hairy, 15 to 30 ft long. Leaves to 10 cm long, palmately lobed with 3 to 7 lobes. Leaf margins toothed. Upper leaf surfaces with short, stiff hairs, rough to the touch. Male and female flowers minute, on separate plants. Fruit to 4 cm long, cone-like, papery (called a *hop*.) Some southwestern tribes used the fruit of the hop to flavor foods and the drink *tulbai*, which is made from fermented maize. Seeds and female flowers were used in bread making. Young shoots have been used as a potherb. The European species *H. lupulus* is utilized in flavoring beer. Found in canyons or along roadsides, particularly near historic ruins.

            AMERICAN HOP
            Mulberry family MORACEAE
            Humulus americanus
            Latin: *humulus*, hop plant

        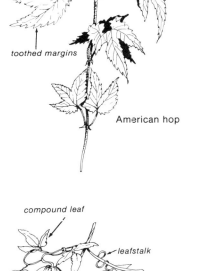
        American hop

        fruit

        3. Leaves compound, leafstalk twisted to aid climbing. Fruit a feathery plume. Flowers showy.

            4. Flower solitary, blue, to 6 cm long, drooping. Leaflets 3 in sets of 2.

                Delicate, beautiful vine. Leafstalks wrap around twigs and branches for support. Leaves to 5 cm, compound, 2 to 3 times parted. Upper surface dull green, lower surface paler. Flowers to 6 cm long with 4 petal-like sepals, purple-blue, lantern-like, in leaf axil. Fruit a feathery plume. Found in canyons and on mountain slopes of the ponderosa pine and mixed conifer forests.

                ROCKY MOUNTAIN CLEMATIS
                Buttercup family RANUNCULACEAE
                Clematis pseudoalpina
                Greek: *klema*, vine; *pseudes*, false

            Rocky Mountain clematis

            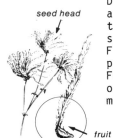
            seed head / fruit

vines

4. Flowers in clusters, white, to 2 cm wide. Leaves pinnate with 3 to 7 leaflets.

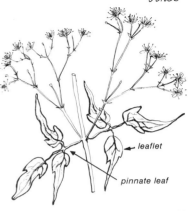
western virgin's bower

Robust vine. Leaves to 8 cm, with 3 to 7 leaflets. Flowers to 1 cm long, white. Male and female flowers on different plants, male flowers more handsome. Fruit a feathery plume. Leaves of this species were chewed by the Indians as a remedy for sore throats and colds. An infusion of the leaves was used for healing cuts and sores on horses. The feathery fruits of both ROCKY MOUNTAIN CLEMATIS and WESTERN VIRGIN'S BOWER make a good tinder for starting fires. Inside their boots hunters can use the fuzz for inner soles and to keep their feet warm. Found along roadsides and in wooded areas of the pinyon-juniper woodland.

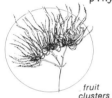
flower

WESTERN VIRGIN'S BOWER
Buttercup family RANUNCULACEAE
Clematis ligusticifolia
Greek: *klema*, vine; Latin: *ligusticum* "of Liguria," an ancient country near modern Genoa; *folium*, leaf

fruit clusters

2. Leaves alternate.

5. Leaves simple, palmately lobed and veined, shiny.

Branches scrambling or weakly climbing. Stems covered with tufts of wool-like hair. Leaves broadly heart-shaped, to 6 cm long, margins coarsely toothed. Leaves sometimes shallowly 3-lobed, both surfaces covered with cottony hairs. Male and female flowers on separate plants. Female plants produce juicy purple-black grapes. All reports of early expeditions to the Rio Grande Valley mention grapes as a food of the Indian pueblos. They ate the fruit either fresh or dried. The raisins were ground and used to flavor food. The grape has been a symbol of revelry and joy throughout literature and art. It is made into wine, grape juice, or fine jellies. Found along streams and banks of the Rio Grande.

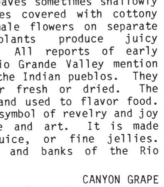
canyon grape

CANYON GRAPE
Grape family VITACEAE
Vitis arizonica
Latin: *vitis*, vine

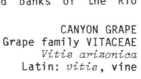
male flowers

5. Leaves palmately compound, 5 to 7 leaflets, leaves drooping.

Woody vine climbing over trees and shrubs. Leaves to 10 cm, dark green, with 5 to 7 leaflets, upper surface shiny, lower covered by a thin down. Tendrils develop disks which adhere to bark and other surfaces. Flowers small, greenish. Fruits to 7 mm, fleshy, blue-black. Found in moist canyons or along roadsides.

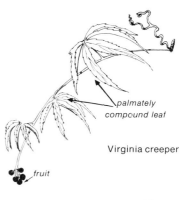

Virginia creeper

VIRGINIA CREEPER
Grape family VITACEAE
Parthenocissus inserta
Greek: *parthenios*, virgin; *kissos*, ivy

1. Stems herbaceous.

    6. Stems leafless, no green color. Parasitic on herbaceous plants. Plants thread-like.

        Twining herbaceous plants with yellow or straw-colored stems. Leaves reduced to minute scales. Flowers to 2 mm, 5-lobed, yellowish to pink. Attached to various herbaceous plants. Found along roadsides. (See also p. 48.)

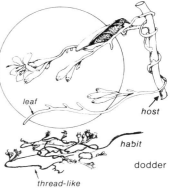

dodder

DODDER
Morning-glory family CONVOLVULACEAE
Cuscuta umbellata
Arabic: *cuscuta*, meaning unknown
Latin: *umbella*, parasol

    6. Stems leafy with green color.

        7. Flowers with colored sepals; no petals present.

            8. Flowers purple. Leafy bract surrounding clusters of three flowers. Leaves opposite, triangular.

                Trailing or clambering plant. Stems to 2 ft long, much-branched, may be in clumps up to 4 ft in diameter. Surfaces sticky, hairy to smooth. Leaves 1 to 5 cm long, somewhat triangular in shape. Bracts forming a leaf-like cup surrounding small purple flowers. Flowers with no petals, sepals purplish red, about 1 cm long. Found in canyons under pinyons or junipers, usually near cliffs.

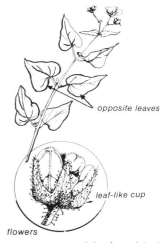

vining four o'clock

VINING FOUR-O'CLOCK
Four-o'clock family NYCTAGINACEAE
Mirabilis oxybaphoides
Latin: *mirabilis*, wonderful
Greek: *oxybaphoides*, like a shallow dish

8. Flowers tiny, greenish white. Fruit triangular. Leaves alternate, heart-shaped.

>Clambering or twining plant. Stems smooth or branched at base. Leaves to 6 cm, heart-shaped with lobes at the base directed backward. Flowers to 4 mm, greenish white, borne in clusters in leaf axils. Fruit 3-angled, black. Often considered a weed. Naturalized from Europe. Found in disturbed areas of the canyons and mesas.

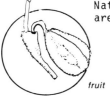
fruit

>BLACK BINDWEED
>Buckwheat family POLYGONACEAE
>*Polygonum convolvulus*
>Greek: *poly*, many, *gonia*, angle
>Latin: *convolvere*, to entwine

7. Flowers with both petals and sepals.

   9. Flowers asymmetrical, 2-lipped.

      >Clambering or trailing plant. Leaves arrowhead-shaped, alternate, surfaces smooth, margins toothed. Flowers to 3 cm long, snapdragoned-shaped, pale blue to rose-red with ridge of yellow-white hairs on lower lip. Seed pods firm and angular. Found in the lower canyons clambering on shrubby species such as New Mexico olive.

seed pod

      >LITTLE SNAPDRAGON VINE
      >Figwort family SCROPHULARIACEAE
      >*Maurandya antirrhiniflora*
      >After a Dr. Maurandy of Cartagena, Spain
      >Greek: *antirrhinos*, against the nose
      >Latin: *flos*, flower

   9. Flowers symmetrical, funnel-like.

      10. Stems trailing. Flowers white or yellow.

          11. Flowers white. Leaves arrowhead-shaped. Leaves and flowers on one side of the stem.

              >A prostrate, trailing plant. Stems smooth to densely hairy. Leaves 2 to 5 cm long, arrowhead-shaped, margins smooth. Flowers funnel-shaped, to 2.5 cm long, single in leaf axils. A pair of leaf-like structures located on stem away from flower. Flowers white to pink with longitudinal dark bands on outside. This plant has been called our worst weed. The root system makes it almost impossible to eradicate once established. When weeded or

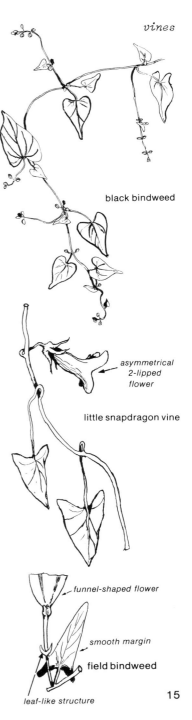
vines

black bindweed

asymmetrical 2-lipped flower

little snapdragon vine

funnel-shaped flower

smooth margin

field bindweed

leaf-like structure

dug, even a small root fragment can begin a new plant. Found along roadsides, in lawns and disturbed sites.

FIELD BINDWEED
Morning glory family CONVOLVULACEAE
Convolvulus arvensis
Latin: *convolvere*, to entwine
arvensis, of cultivated fields

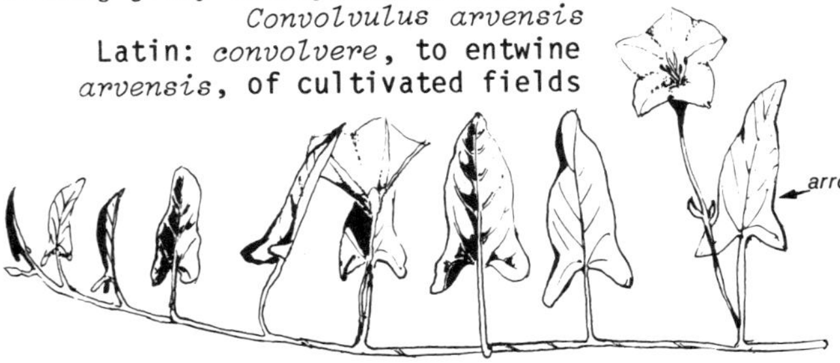

field bindweed

arrowhead-shaped leaves

11. Flowers yellow. Leaves large, to 1 ft long, triangular.

A large ill-smelling plant which may grow to 10 ft across. Leaves to 1 ft long, somewhat triangular. Margins toothed to smooth. Flowers to 15 cm long, either male or female; anthers often twisted in male flowers. Fruit to 7 cm across, yellow when ripe. Found along roadsides and in disturbed soils along riverbanks. (See also p. 135.)

fruit

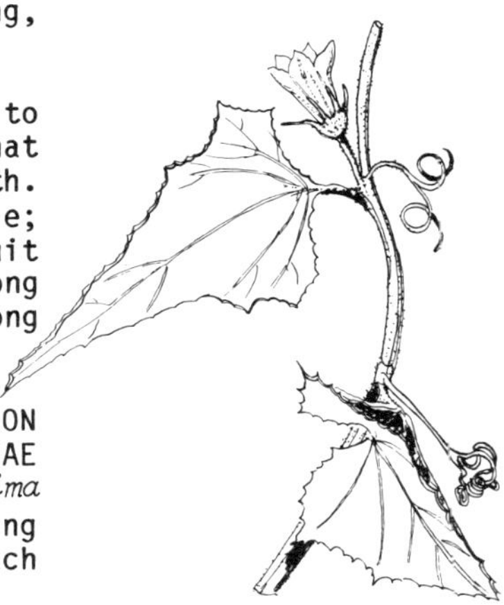

COYOTE MELON
Gourd family CUCURBITACEAE
Cucurbita foetidissima
Latin: *cucurbita*, gourd; *foetidus*, stinking
-issima, very much

coyote melon

10. Stems twining. Flowers red. Young leaves deeply lobed.

Herbaceous plant twining on vegetation. Leaves to 5 cm, lobed, margins smooth. Flowers to 4 cm, bright red, funnel-shaped. Stamens project beyond flower (exerted). Sepals in 2 series, the outer series larger than the inner. Found along roadsides and disturbed areas of canyons and pinyon-juniper woodland.

stamens

sepals

flower

lobed leaf

STAR-GLORY
Morning glory family CONVOLVULACEAE
Ipomoea coccinea
Greek: *ipos*, worm; *kokkinos*, scarlet

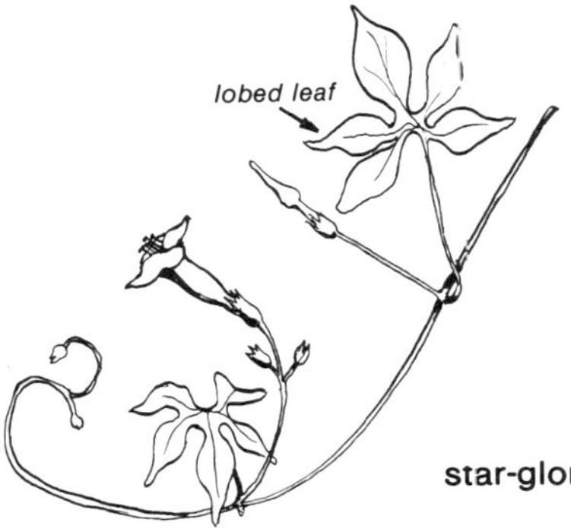

star-glory

TREES

Trees are woody plants with one (or a few) main stem(s) called a *trunk*. They grow 13 to 160 feet or more. Many trees are long-lived. Large pines can live 200 to 300 years. Trees may be *evergreen*, retaining some leaves throughout the year, or *deciduous*, losing their leaves every fall.

This chapter has four sections:

     I. LEAVES NEEDLE-LIKE, EVERGREEN, p. 17
    II. LEAVES SMALL, TRIANGULAR, SCALE-LIKE, p. 20
   III. LEAVES DECIDUOUS, COMPOUND, p. 22
   IV. LEAVES DECIDUOUS, SIMPLE, p. 23

I. LEAVES NEEDLE-LIKE.

1. Needles attached to twig in bundles of 2 to 5.

  2. Needles in bundles of 5.

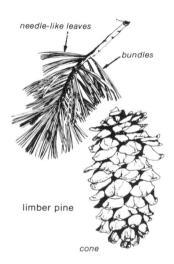

needle-like leaves
bundles
limber pine
cone

habit

A moderately large tree, to 60 ft tall, with drooping, plume-like branches. Branches flexible, as indicated by the specific name *flexilis*. Twigs orange when immature, becoming gray with age. Needles green, to 6 cm long, with minute teeth on the margins, clustered at ends of the branches. Found scattered throughout the mixed conifer forest.

                       LIMBER PINE
                Pine family PINACEAE
                    Pinus flexilis
    Latin: *pinus*, pine; *flectere*, to bend

  2. Needles in bundles of 2 to 3.

    3. Tree with much-branched trunk. Seeds nut-like. Needles in bundles of 2, occasionally single.

pinyon pine
seed

Bushy, evergreen tree, to 35 ft tall. Needle-like leaves usually two to a bundle. The top-shaped cones open in September, releasing nut-like edible seeds. The pinyon has been important in both the Spanish and Indian cultures of the Southwest. The nuts are edible, the pitch used for waterproofing and repairing pottery vessels, and the gum used medicinally. Dominant tree of the pinyon-juniper woodland; generally found at altitudes 6500 to 7500 ft.

habit

                       PINYON PINE
                Pine family PINACEAE
                     Pinus edulis
            Latin: *edere*, to eat

3. Tall tree with single erect trunk. Needles in bundles of 3.

habit

Tall tree up to 230 ft; 5 to 8 ft in diameter. Branches stout, thick and somewhat drooping. Older bark yellow-brown with a vanilla odor. Cones 7 to 13 cm long, red-brown with prickly scales; seeds winged. Both Indians and pioneers used this tree for building material and firewood. In times of starvation, the Indians harvested the inner bark for food. This pine is an important food source for many animals, including squirrels, chipmunks, and a number of species of birds. Occurs in nearly pure stands at altitudes of 7000 to 8000 ft; along streams at lower elevations. Scattered throughout the mixed conifer forest at higher elevations.

PONDEROSA PINE
Pine family PINACEAE
Pinus ponderosa
Latin: *pondus*, weight

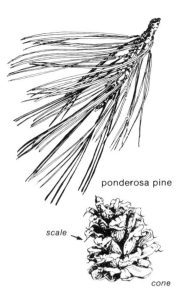

ponderosa pine

scale

cone

1. Needles attached singly to the twig.

4. Needles flat.

5. Needles attached by small stems. Cones with sharp 3-pointed bracts protruding from between the scales.

habit

A tall tree reaching a height of 190 ft and a diameter of 3 to 4 ft. Branches slender, long, and drooping. Needle-like leaves short-stalked, blue-green, soft, flat, arranged spirally on the branches. Cones red-brown, hanging, with distinctive 3-parted bracts. Bark reddish to gray-brown. These trees provide cover for wildlife, and seeds are eaten by squirrels and other rodents. Found scattered throughout the mixed conifer forest and along north-facing slopes in the canyons.

DOUGLAS-FIR
Pine family PINACEAE
Pseudotsuga menziesii
Greek: *pseudes*, false
Japanese: *tsuga*, hemlock tree
Menziesii honors Scottish naturalist
Archibald Menzies (1754-1842)

bract

Douglas-fir

5. Needles attached by disk. Buds blunt and pitch-covered. Cones erect, in upper branches, disintegrating on the tree.

   6. Current season's twigs without hairs. Cones yellow or gray-green. Needles blunt. Bark firm.

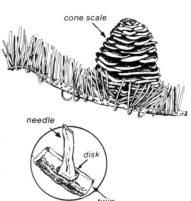

trees

cone scale

needle

disk

twig

white fir

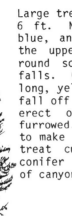

habit

Large tree, to 250 ft tall, with diameter 3 to 6 ft. Needle-like leaves flat, blunt, pale blue, and attached singly; needles erect on the upper surface of the twigs. A small, round scar remaining on twig after needle falls. Cones erect, oval-shaped, 8 to 13 cm long, yellow to purple in color. Cone scales fall off leaving central axis of cone standing erect on branch. Bark gray and deeply furrowed. New Mexico Indians used the twigs to make pipe stems and the resinous pitch to treat cuts. Commonly found in the mixed conifer forest or on the north-facing slopes of canyons.

                                      WHITE FIR
                              Pine family PINACEAE
                                  Abies concolor
                    Latin: *abies*, fir-tree
                      concolor, single color

   6. Current season's twigs with hairs. Cones dark purplish brown. Needles blunt. Bark spongy.

bark

corkbark fir

Tall tree to 90 ft and 3 ft in diameter, generally smaller than white fir. Bark ashy gray, not breaking into plates. Twigs soft-hairy. Much less common than the white fir. Scattered throughout the higher elevations.

                                  CORKBARK FIR
                              Pine family PINACEAE
                                  Abies lasiocarpa
        Greek: *lasios*, shaggy; *karpos*, fruit

4. Needles 4-angled, sharp-pointed, rigid. (Needles can be twirled in fingers). Twigs from which needles have fallen bumpy.

   7. Young twigs without hairs. (Use hand lens.) Needles very stiff, uncomfortable to grasp. Cone over 6 cm long.

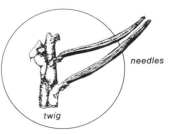

needles

twig

Colorado blue spruce

cone

Tree to 80 ft, 2 ft in diameter, with a narrow conical shape. Bark gray to reddish brown and divided into ridges. Leaves stiff and sharp, covered with a white, waxy coating. Found throughout the mixed conifer forest, rarely in pure stands.

COLORADO BLUE SPRUCE
Pine family PINACEAE
Picea pungens
Latin: *picea*, spruce-tree
pungere, to puncture

7. Young twigs with hairs. Needles slightly pliable, not sharp. Cone less than 6 cm.

Tree to 120 ft tall and 3 ft in diameter. Needles dark bluish green but not stiff. Bark cinnamon-red, separated into scales. Common spruce in the spruce-fir forest.

ENGELMANN SPRUCE
Pine family PINACEAE
Picea engelmannii
In honor of George Engelmann, German-born physician and naturalist (1809-1884)

Engelmann spruce

II. LEAVES SMALL, TRIANGULAR, SCALE-LIKE.

1. Clusters of pink flowers at ends of branches in spring and summer. Branches delicate, wiry, drooping.

   Small tree or shrub to 20 ft, with scale-like leaves which fall in the winter. Branches upright, spreading. Leaves blue-green. Considered a weed tree which competes with and replaces native vegetation along water-holding areas and rivers. Imported from the Mediterranean in 1823, it consumes large amounts of water through evapotranspiration, diminishing already scarce western water supplies. Found along the banks of the Rio Grande and occasionally roadsides.

   ### TAMARISK, SALT-CEDAR
   ### Tamarisk family TAMARICACEAE
   Tamarix pentandra
   Arabic: *tamr*, a date. Greek: *pente*, five
   andros, male, i.e., five-stamened

1. No showy flowers. Branches stout.

   2. Bark of main trunk thick, checkered into squarish plates. Fruit usually 4-seeded.

      Tree to 40 ft tall, with large trunk. Bark of the younger branches smooth, reddish beneath. Older bark divided into rectangular or

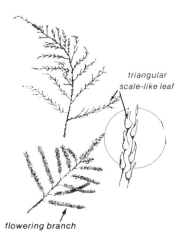
tamarisk, salt-cedar

trees

squarish plates like the scales of an alligator. Leaves blue-green with white dots. Confined to dry slopes 6500 to 8000 ft.

ALLIGATOR JUNIPER
Cypress family CUPRESSACEAE
Juniperus deppeana
Latin: *iuvenis*, youth; *papere*, to produce
In honor of Ferdinand Deppe, German botanist
(d. 1861)

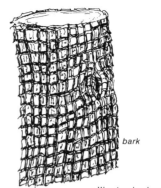

bark

alligator juniper

2. Bark shredding, fibrous.

   3. Branchlets flattened, drooping, silver-green. Mature trees to 50 ft.

      Tree to 50 ft tall, with branches flattened, silvery in color, drooping gracefully. Fruit with 1 to 2 seeds. Bark reddish brown to gray. Pueblo Indians reportedly ate the berries either raw or stewed. Junipers provide important food and cover for wildlife. Found in canyons.

ROCKY MOUNTAIN JUNIPER
Cypress family CUPRESSACEAE
Juniperus scopulorum
Latin: *scopulae*, broom

   3. Branchlets rigid, upright, olive green. Small tree, not over 25 ft, usually with multiple trunks, often with many dead branches.

      Much-branched scraggly tree to 25 ft tall, with yellow-green scale-like leaves. Fruit dark blue to brown with one seed. Bark gray, fibrous, and shreddy. Male and female flowers on separate trees, the female tree producing a berry-like fruit. Junipers have been widely used for firewood and fence posts by New Mexicans. The shreddy bark was bound into compact bundles with cord made of yucca fibers and then used as a torch to light dwellings of the early Tewa Indians. It was also used for mats, saddles, breechcloths, and cradle boards. Common co-dominant species of the pinyon-juniper woodland.

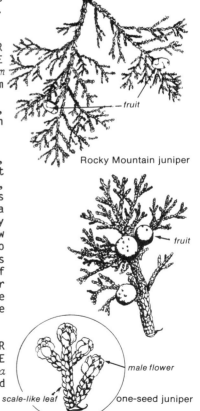

fruit

Rocky Mountain juniper

fruit

ONE-SEED JUNIPER
Cypress family CUPRESSACEAE
Juniperus monosperma
Greek: *monos*, single; *sperma*, seed

habit

scale-like leaf

male flower

one-seed juniper

III. LEAVES DECIDUOUS, COMPOUND.

1. Leaves pinnately compound.

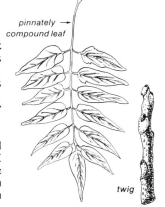

pinnately compound leaf

twig

tree-of-heaven

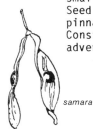

samara

Medium-size tree which may grow to 100 ft tall. Spreads by root sprouting. Flowers small, yellow-green, unpleasant-smelling. Seed a twisted fruit called a *samara*. Leaves pinnately compound with 11 to 25 leaflets. Considered a weed tree. Grows well under adverse conditions. Sometimes in canyons.

TREE-OF-HEAVEN
Quassia family SIMAROUBACEAE
Ailanthus altissima
Malaccan: *ailanto*, tree-of-heaven
Latin: *altis*, high

1. Leaves palmately compound.

   2. Leaves opposite, leaflets rather broad.

palmately compound leaf

boxelder maple

samaras

Tree growing to 75 ft with a broad, rounded crown. Bark pale gray to brown, divided into narrow ridges. Leaves 15 cm long with 3 leaflets. Margins saw-toothed. Two-winged seed pods (samaras) droop from branches. Tewa Indians used the twigs for making pipe stems; the Mescalero Apache boiled the inner bark and outer portions beneath the bark until the sugar crystallized. Seeds provide nourishment for birds and squirrels; foliage provides nesting areas. Grows along streams.

BOXELDER MAPLE
Maple family ACERACEAE
Acer negundo
Latin: *acer*, sharp (refers to leaves)
Foliage resembles that of *Vitex negundo*,
an unrelated Malayan tree

   2. Leaves alternate, leaflets rather narrow.

alternate leaf

narrowleaf hoptree

A small tree or shrub to 10 ft tall with compound leaves. Leaflets 3, dark green; when crushed, exude a skunk-like odor. Flowers pale yellow to greenish white, fragrant. Fruit coin-shaped, remaining on tree after leaves fall. Fruits have been used as a substitute for hops in beer making. Roots at one time were used for quinine in the prevention of malaria. Some people will get a dermatitis when they come in contact with the tree. Found mostly in canyons of the pinyon-juniper woodland and ponderosa pine forest.

trees

NARROWLEAF HOPTREE
Rue family RUTACEAE
Ptelea trifoliata
Greek: *ptelea*, elm tree
Latin: *tri*, three; *foliate*, leaf

IV. LEAVES DECIDUOUS, SIMPLE.

1. Leaves opposite.

    2. Leaves lobed with pointed teeth or deeply dissected.

        A small tree or large shrub to 25 ft, with simple or palmately compound leaves, 3- to 5-lobed. Upper leaf surface dark green, lower pale. Indians made an excellent bow from the branches. Seeds and flowers provide food for a number of species of birds, squirrels, and chipmunks. Found along stream banks or in moist areas throughout the ponderosa pine and mixed conifer forests.

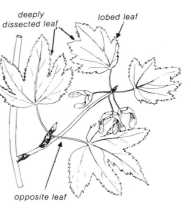
Rocky Mountain maple

ROCKY MOUNTAIN MAPLE
Maple family ACERACEAE
Acer glabrum
Latin: *glaber*, bald

   2. Leaf margins smooth, not lobed.

        Thicket-forming small tree or shrub to 16 ft tall. Bark dark red with prominent dots (lenticels). Branches reddish or purplish, smooth and slender. Flowers in flat-topped clusters, each flower with a 4-lobed sepal. Petals dull white. Fruit dull white, pea-shaped. Found in canyon bottoms.

RED-OSIER DOGWOOD
Dogwood family CORNACEAE
Cornus stolonifera
Latin: *cornus*, dogwood
stolo, a shoot; *ferre*, to bear

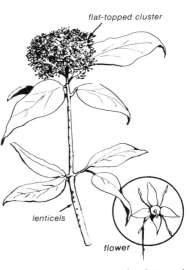
red-osier dogwood

1. Leaves alternate.

    3. Leaves lobed. Fruit an acorn.

        A large shrub or tree growing to 70 ft tall, often in thickets. Trunk to 4 ft in diameter. Leaves oblong, to 17 cm long; lobed halfway to the midrib vein. Upper surfaces dark green, undersides light green, softly hairy. Leaves turning bronze in the fall, remaining on the branches throughout the winter. The wood was used by various Indian tribes for making implements. The acorns were an important food source for both Indian and wildlife such as wild turkey. Twigs and foliage provide browse for deer and elk. Found throughout the area. (See p. 44.)

acorn

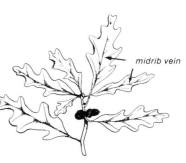
midrib vein

Gambel's oak

GAMBEL OAK
Beech family FAGACEAE
Quercus gambelii
Latin: *quercus*, oak. *Gambelii* honors William Gambel (1819-1849), American naturalist

3. Leaves not lobed.

    4. Leafstalk long, flattened. Leaves flutter.

        5. Leaves nearly circular. Bark whitish. Trees often in dense groves.

Tree to 80 ft tall, with distinctive smooth, white to grayish bark. Flowers in drooping clusters called *catkins*. Seeds cottony. Leaves somewhat oval with a pointed tip. Leafstalk flattened so leaf trembles. Bark was used extensively by pioneers and Indians as a remedy for fevers and for prevention of scurvy. Contains salicin and populin, which are relatives of aspirin. Aspens are highly susceptible to fire damage and when burned sprout from the root. Found at higher elevations; pioneers on burned or logged areas.

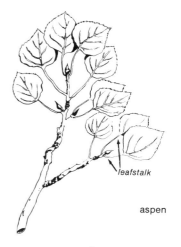
leafstalk

aspen

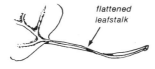
flattened leafstalk

ASPEN
Willow family SALICACEAE
Populus tremuloides
Latin: *populus*, poplar tree
tremere, to tremble

        5. Leaves roughly triangular, wider than long. Bark gray. Plants of stream banks.

Large, rapidly growing tree which may reach 90 ft. Leaves broadly oval to triangular, yellow-green and smooth. Male and female flowers on separate trees. Female trees produce seeds covered with cottony hairs. This tree has been used for fuel and fence posts. To the early settler the cottonwood meant shade and water. Cottonwood is a

Rio Grande cottonwood

favorite drum wood of the pueblos and the Hopi use it for kachinas. Found along streams and in moist areas.

RIO GRANDE COTTONWOOD
Willow family SALICACEAE
Populus fremontii
Honors John C. Fremont (1813-1819), explorer

4. Leafstalk (petiole) rounded.

  6. Old wood stems thorny.

    7. Leaves long and narrow, underside silvery. Fruit silvery.

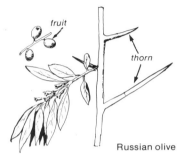

Russian olive

Tree to 25 ft tall. Leaves narrow, with silver dots on the underside. Flowers pale yellow, bell-shaped, strong-smelling. Fruit short-stemmed, spherical, silvery. Introduced into the United States as a shade tree, now an escape. Found along watercourses such as the Rio Grande.

bell-shaped flower

RUSSIAN OLIVE
Oleaster family ELAEAGNACEAE
Elaeagnus angustifolia
Greek: *elaion*, olive oil; *eygos*, willow
Latin: *angustus*, narrow; *folium*, leaf

    7. Leaves elliptical, toothed or lobed. Twigs gray, thorns smooth and tapering. Fruit yellow, red, or black.

hawthorn

Tree to 10 ft tall. Branches spiny. Leaves simple, toothed or shallowly lobed. Flowers white with 5 sepals, 5 petals. Fruits egg-shaped, usually red. Found in canyon bottoms. (See p. 37.)

fruit

HAWTHORN
Rose family ROSACEAE
Crataegus erythropoda
Greek: *kratos*, strength
erythros, red; *podos*, foot

  6. Stems without thorns.

    8. Leaves narrow, 4 times longer than wide.

Tree growing to 60 ft tall, trunk to 2 ft in diameter. Bark gray, irregular, and narrowly fissured. Leaves willow-like, lance-shaped with margins slightly toothed. Upper leaf surface light green and smooth, lower lighter and hairy. Male and female flowers in drooping catkins that appear before the leaves. Female trees produce a cottony seed. Indians of the Southwest ate uncooked fruits and the inner bark to prevent scurvy. Shoots were used by various tribes for basket weaving. Found along streams and in moist areas.

catkin

NARROWLEAF COTTONWOOD
Willow family SALICACEAE
Populus angustifolia
Latin: *populus*, poplar tree
angustus, narrow; *folium*, leaf

8. Leaves not longer than wide.

   9. Leaves doubly saw-toothed.

      10. Catkin scales woody and persistent. Pith of twigs triangular in cross section.

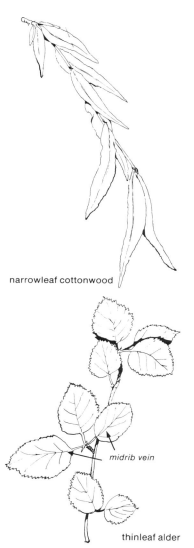

narrowleaf cottonwood

Tree to 30 ft tall. Leaves oval to oblong with heart-shaped bases and pointed tips, dark green above, pale green or yellow below. Midrib vein (large vein going down the center of the leaf) orange-colored. Leafstalk (petiole) stout, slightly grooved and orange-colored. Twigs covered with soft hairs and marked by large, orange-colored dots (lenticels). When freshly cut the white bark turns pink-red. Female flowers in cone-like catkins; male flowers in elongated hanging clusters called *aments*. The Tewa Indians used the bark of alder alone or with other plants to form a red dye for coloring or painting buckskin. Grows along streams.

female

catkin

male

midrib vein

THINLEAF ALDER
Birch family BETULACEAE
Alnus tenuifolia
Latin: *alnus*, alder tree
tenuis, slender; *folium*, leaf

thinleaf alder

      10. Catkin scales thin and deciduous. Pith of twigs round in cross section.

Tree to 25 ft tall. Bark black on young trees, turning reddish brown with conspicuous white horizontal dots (lenticels). Leaves broadest at base, tip pointed, dark greenish yellow and shiny above, paler below and dotted. Male and female flowers in catkins.

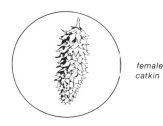
female catkin

Conelets 3 cm long with 3-lobed scales. Found along streams.

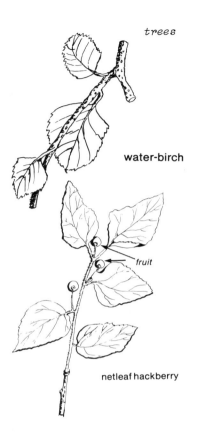

trees

water-birch

fruit

netleaf hackberry

WATER-BIRCH
Birch family BETULACEAE
Betula occidentalis
Latin: *betula*, birch tree; *occidere*, to fall
(i.e., of the west, where the sun sets)

9. Leaves with smooth or toothed margins but not doubly saw-toothed. Orange spherical fruit in leaf axils.

Tree to 30 ft tall. Bark grayish brown to dark brown, covered with wart-like projections. Leaves broadest near base, tapering to a point, uneven at base, toothed on the margins midway to tip, veins conspicuous. Fruit fleshy, spherical, orange-red to red-brown. Along banks of the Rio Grande and arroyos.

NETLEAF HACKBERRY
Elm family ULMACEAE
Celtis reticulata
Latin: *celtis*, Pliny's name for an African species of lotus; *reticulum*, a fine net

SHRUBS

Shrubs are woody species which have multiple stems. They usually are less than 16 feet tall. Some are sprawling or creeping, others erect. Many shrubs sprout after fires to form thickets from underground root systems. They provide valuable food and shelter for wildlife.

This chapter is divided into four keys based on habit and position of the leaves:

    I. LOW, CREEPING UNDERSHRUBS LESS THAN 15 CM TALL, p. 28
   II. SHRUBS WITH MATURE GROWTH OVER 15 CM; LEAFLESS OR LEAVES IN BASAL ROSETTES OR NEEDLE-LIKE, p. 29
  III. SHRUBS WITH MATURE GROWTH OVER 15 CM; LEAVES COMPOUND, p. 30
  IV. SHRUBS WITH MATURE GROWTH OVER 15 CM; LEAVES SIMPLE, p. 35

I. LOW, CREEPING UNDERSHRUBS LESS THAN 15 CM TALL.

1. Leaves compound, strongly spine-toothed, holly-like, shiny.

    Low, creeping, evergreen shrub growing to 30 cm high. Stems lying on ground. Leaves alternate, pinnately compound with 3 to 7 leaflets. Leaflets oval to oblong with spiny-tipped lobes, holly-like in appearance. Flowers yellow, 1 to 3 cm wide. Fruit a berry, bluish, in clusters. Barberry purportedly was used by Indians for the treatment of rheumatism and fevers. Spanish New Mexicans, who used it to treat anemia, called it *yerba de la sangre*. Also called CREEPING MAHONIA. Shady slopes, mixed conifer forest.

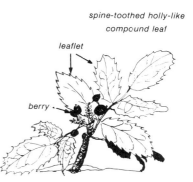

creeping barberry

flower

CREEPING BARBERRY
Barberry family BERBERIDACEAE
Berberis repens
Arabic: *berberys*, barberry fruit
Latin: *reptare*, to crawl

1. Leaves simple, not spine-toothed.

    2. Leaves alternate.

        3. Leaves somewhat leathery, margins smooth.

            Low, mat-forming evergreen shrub. Leaves shiny green, leathery. Flowers urn-shaped, white to pink, 6 mm wide. Fruits red, in

kinnikinnik
flower

clusters. Early settlers and Indians called the plant *coralillo* and used the leaves as a substitute for tobacco. It was pulverized, carried in pouches, and became known as *knick-knack*, giving the common name, *kinnikinnik*. Forms carpets beneath pines of ponderosa and mixed conifer forests.

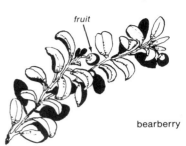
bearberry

habit

BEARBERRY, KINNIKINNIK
Heath family ERICACEAE
Arctostaphylos uva-ursi
Greek: *arktos*, bear; *staphyle*, bunch of grapes; Latin: *uva*, grape; *ursus*, bear

3. Leaves not leathery, margins toothed. Stems brown to gray, 4-angled (square) in cross section.

Low shrub with strongly angled branches. Leaves oval with toothed margins. Petals to 4 mm long, somewhat urn-shaped, white to rose. Fruit blue with a white covering. Common ground cover in mixed conifer and spruce-fir forests.

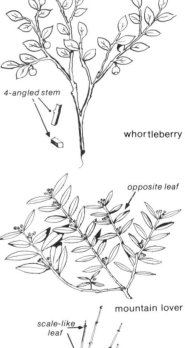
whortleberry

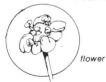
urn-shaped flower

WHORTLEBERRY
Heath family ERICACEAE
Vaccinium myrtillus
Greek: *vacca*, cow; Persian: *murd*, myrtle

2. Leaves opposite, shiny, margins saw-toothed.

Leaves oval to elliptical with margins rolled under. Flowers tiny, to 2 mm, green-brown or dark red, 4-pointed, in axils of leaves. Found in ponderosa and mixed conifer forests.

mountain lover

MOUNTAIN LOVER
Staff-tree family CELASTRACEAE
Pachystima myrsinites
Greek: *pachys*, thick; *stime*, stigma; Persian: *murd*, myrtle

flower

II. SHRUBS WITH MATURE GROWTH OVER 15 CM; LEAFLESS OR LEAVES IN BASAL ROSETTES OR NEEDLE-LIKE.

1. Leafless shrub. Stems smooth and jointed.

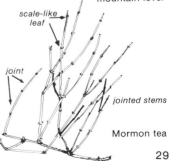
Mormon tea

cone-like flower

Evergreen shrub with jointed stems. Leaves scale-like, inconspicuous. Flowers minute, cone-like in axils of branchlets. Male and female flowers on separate plants. Indians and Spanish used the stems to make a tea. The common name MORMON TEA or BRIGHAM TEA commemorates use of the plant by early Mormons. On dry slopes of the pinyon-juniper woodland.

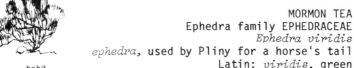
habit

MORMON TEA
Ephedra family EPHEDRACEAE
Ephedra viridis
ephedra, used by Pliny for a horse's tail
Latin: *viridis,* green

1. Shrub with obvious leaves.

    2. Leaves small, needle-like, 1.5 cm long, white line on upper surface.

        Low, spreading shrub no more than 3 ft in height. Leaves needle-like, 1.5 cm long. Upper surface grooved with a white line, underside shiny green. Fruit dark blue, with a waxy covering. All parts of the plant produce a volatile oil which has been used as a diuretic. Oils of the berry give gin its distinctive flavor and are also used in the manufacture of varnish. Found in the mixed conifer forest.

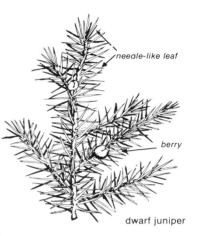

DWARF JUNIPER
Cypress family CUPRESSACEAE
Juniperus communis
Latin: *iuvenis,* youth; *parere,* to produce
communis, common

habit

    2. Leaves very large, stiff, spine-tipped, to 3 ft long, fleshy, all attached at base of plant (See p. 101-102.)

YUCCA
Lily family LILIACEAE
Yucca spp.
West Indian: *yuca,* a *Yucca* plant

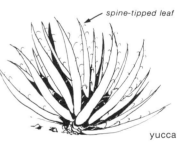

III. SHRUBS WITH MATURE GROWTH OVER 15 CM; LEAVES COMPOUND.

1. Stems and/or branches with bristles, prickles, or spines. Leaves pinnately compound.

    2. Twigs with large, sharp spines. Flowers pink, in large clusters. Fruit a legume.

flower

Thicket-forming shrubs. Branches with straight or curved spines. Leaves pinnately compound, blue-green, somewhat hairy. Poisoning has been reported from eating bark, roots, and seeds, but the New Mexican Indians

gathered the flowers, boiled them, then ate them as a vegetable. The tough, elastic wood made choice bows. Found in canyons, along roadsides and trails of the pinyon-juniper woodland and ponderosa pine forest.

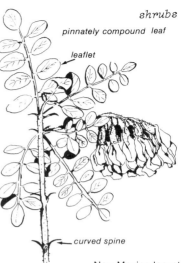

shrubs

pinnately compound leaf

leaflet

curved spine

New Mexico locust

NEW MEXICO LOCUST
Pea family LEGUMINOSAE
Robinia neomexicana
Honors Jean Robin (1550-1629)
French herbalist

2. Stems usually with prickles or bristles.

   3. Leaflets 3 to 5, coarsely toothed. Stems bristly. Thicket-forming.

      Shrubs growing to 3 ft. Compound leaves with 3 to 5 leaflets. Undersurface white, margins sharply toothed. Flowers white, in small clusters at end of stem. Petals 5, stamens numerous. Fruit bright red. Raspberries are pioneer plants on recently burned or disturbed areas. An important source of food and shelter for birds and small mammals. Canyons and mountain slopes of high elevations.

berry

WILD RASPBERRY
Rose family ROSACEAE
Rubus strigosus var. *arizonicus*
Latin: *rubeus*, red; *striga*, furrow

bristle

wild raspberry

   3. Leaflets 5 to 15, evenly toothed. Stems prickly.

WILD ROSE
Rose family ROSACEAE
Rosa spp.
Latin: *rosa*, a rose

Several species of wild rose grow on the Pajarito Plateau.

The edible fruit, called a *hip*, is dark red when ripe. Rose hips have high concentrations of vitamins C and A. They have been used medicinally and as a tea. Spanish New Mexicans collected, dried, and ground rose petals for use as a sore-throat remedy. Members of the genus hybridize freely and the identification of species is often difficult. The following varieties of *Rosa woodsii* have been found in the area:

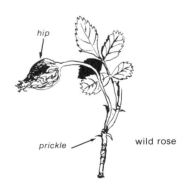

hip

prickle

wild rose

FENDLER'S ROSE, *R. woodsii var. fendleri*. Prickles straight, sparse. Flowers pink, solitary or in clusters of 2 or 3, fruit round.

PALELEAF ROSE, *R. woodsii var. hypoleuca*. Prickles straight, numerous. Flowers pink, in flat-topped clusters at ends of stems. Canyons and mountain slopes.

Fendler's rose

1. Stems and/or branches without bristles, prickles, or thorns.

    4. Leaves with 3 leaflets (palmately compound).

        5. Leaflets less than 4 cm each, lobed. Strong-smelling when bruised. Berries orange-red, in clusters.

            Shrub growing 3 to 10 ft, often in thickets. Leaves compound with 3 leaflets. Fruit orange-red. The stems were used by several Indian tribes as one of the chief materials in basket-making. Dyes for leather and baskets have been made from the leaves. Stems were used for bows and arrowshafts. The fruit is eaten by a number of species of birds; foliage is browsed by deer. Also called SQUAWBUSH, LEMONADE BERRY, LEMITA. In deep soils of canyons and mesas.

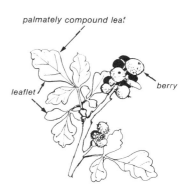
skunkbush

SKUNKBUSH
Sumac family ANACARDIACEAE
Rhus trilobata
Greek: *rhous*, sumac
Latin: *trilobatus*, three-lobed

   5. Leaflets over 4 cm.

      6. A small tree with leaves having a skunk-like odor.

         Small tree or large shrub seldom over 6 ft. Leaves dark green, compound, with 3 leaflets, foul-smelling. Flowers tiny, to 7 mm, greenish. Fruit flattened with disk-like wings, surface netted. Canyons and rocky places of the pinyon-juniper woodland and ponderosa pine forest. (See also p. 22.)

narrowleaf hoptree

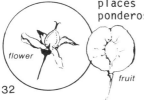

NARROWLEAF HOPTREE
Rue family RUTACEAE
Ptelea trifoliata
Greek: *ptelea*, elm
Latin: *tri*, three; *foliate*, leaf

shrubs

6. A low-growing undershrub with white berries. Leaves smooth, drooping. CAUTION: Do not touch.

> Single-stemmed shrub with compound leaves. Leaflets bright green, turning orange to red in the autumn. Margins coarsely toothed. Flowers small, greenish yellow. Berries white, dull, waxy, and smooth. About half the human population is susceptible to the non-volatile oils in this plant. In these persons it causes a dermatitis. Berries are toxic to man but are eaten by 75 species of birds. Along streams in canyons.
>
> POISON IVY
Sumac family ANACARDIACEAE
Rhus radicans
Latin: *radix*, root

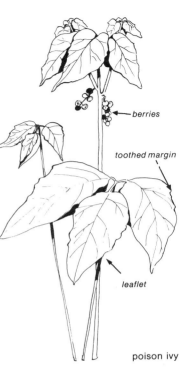

poison ivy

4. Leaves with more than 3 leaflets (pinnately compound).

   7. Leaves opposite.

      8. Leaves once-pinnately compound. Flowers small, white, in terminal clusters. Fruit red, juicy.

         > Shrub growing up to 24 ft. Young twigs hairy, mature twigs smooth; red to yellow-brown. Leaves compound, 5 to 7 leaflets per leaf, oval to lance-shaped with margins sharply toothed. Flowers fragrant, creamy white, in pyramidal clusters. Fruit red, juicy. Indians of California called elderberry the "tree of music" because flutes were made from the straight stems. There are reports, however, that children have been poisoned when they used blowguns made from the bark. Red elderberries can be toxic, especially if eaten raw. Elderberry wine is made from the related black elderberry *S. melanocarpa*. Found in shaded canyons.

fruit

RED ELDERBERRY
Honeysuckle family CAPRIFOLIACEAE
Sambucus microbotrys
Latin: *sambucus*, an elder tree
Greek: *mikros*, small
botrys, a bunch of grapes

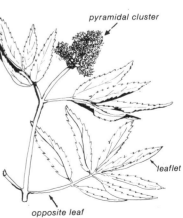

red elderberry

   8. Leaves thrice-pinnately compound. Berries white or red in long clusters at top of stalk.

      > Shrub-like plant growing to 3 ft tall. Stems smooth to hairy. Leaflets oval, 3-to 5-lobed or coarsely toothed. Flowers small, white, in

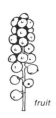

fruit

dense long clusters at the end of the stem. This plant is poisonous to man. Berries contain alkaloids which act upon the heart; death may ensue. A good rule of thumb is to avoid berries which are red or white. In canyons of mixed conifer forest.

WESTERN BANEBERRY
Buttercup family RANUNCULACEAE
Actaea arguta
Greek: *aktaea*, an elder tree
Latin: *argutus*, sharp of taste

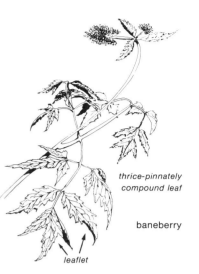

thrice-pinnately compound leaf

baneberry

leaflet

7. Leaves alternate.

   9. Leaflets 5, linear, silky. Flowers yellow. Bark shreddy.

Shrub growing to 4 ft. Twigs silky-hairy when young, later becoming shreddy. Compound leaves with 3 to 7 leaflets, silky-hairy, margins inrolled, evergreen. Petals 5, heart-shaped, 5 bracts alternating with sepals. The only woody potentilla in New Mexico. Often used as an ornamental in southwestern gardens. High meadows.

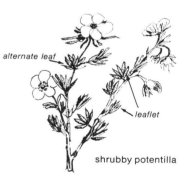

alternate leaf

leaflet

shrubby potentilla

sepals

bracts

flower

leaflet inrolled margins

SHRUBBY POTENTILLA
Rose family ROSACEAE
Potentilla fruticosa
Latin: *potens*, powerful; *fructus*, fruit

   9. Flowers purple, fruit like a pea-pod (legume).

      10. Small, low shrub of rocky areas. Twigs light gray, pointed. Leaves glandular-dotted. Sepals feathery.

legume

Shrub growing to no more than 3 ft tall. Leaves small, oddly pinnate, glandular dots on lower surface. Flowers pea-like, rose-pink to yellow. Bracts below flowers silky and feather-like on margins. Sepals densely hairy. Fruits flat, hairy. Plants have been used by Spanish New Mexicans for treating rheumatism. Rocky slopes near Rio Grande.

feathery sepals

pea-like flower

FEATHER INDIGOBUSH
Pea family LEGUMINOSAE
Dalea formosa
Honors Samuel Dale (1659-1739)
English physician and botanist
Latin: *formosus*, beautiful

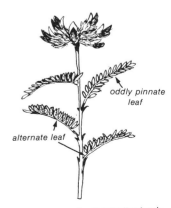

oddly pinnate leaf

alternate leaf

feather indigobush

10. Tall shrubs of stream banks. Flowers with 1 petal. Fruit pea-like, dotted.

>Shrubs over 6 ft tall. Leaflets elliptical to oval, with dots. Flowers in dense spikes, purple. Sepals hairy. Along banks of Rio Grande.

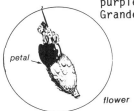

>>FALSE INDIGOBUSH
>>Pea family LEGUMINOSAE
>>*Amorpha fruticosa*
>>Greek: *amorphos*, irregular
>>Latin: *fructus*, fruit

false indigobush

IV. SHRUBS WITH MATURE GROWTH OVER 15 CM; LEAVES SIMPLE.

1. Stems or twigs with spines.

    2. Leaves in a circle around the stem or in bundles at the nodes.

        3. Spreading shrub less than 2 ft high. Spines along branches.

>Thicket-forming shrubs. Leaves evergreen, broadly elliptical with smooth margins, undersides hairy. Flowers white, in clusters at the ends of the twigs. Stamens extending beyond the petals. Many species of *Ceanothus* contain saponin, which gives the fruits and flowers soap-like qualities. Indians and early settlers used the crushed flowers for soap. Throughout burned areas and along trails and roads. (See also p. 37)

>>BUCKBRUSH
>>Coffeeberry family RHAMNACEAE
>>*Ceanothus fendleri*
>>Greek: *keanothos*, a kind of thistle
>>Honors August Fendler (1813-1883)
>>German-born naturalist and explorer

buckbrush

3. Upright shrub. Spines at the nodes.

    4. Irregularly branched shrub. Leaves gray-green with smooth margins, single spine at each node.

>Shrub to 4 ft. Twigs gray and spiny. Leaves arranged in bundles around stem. Flowers to 5 cm, funnel-shaped, green-yellow with purple veins. Stamens extending beyond petals. Berries spherical, scarlet, resembling tiny

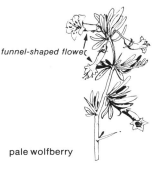

pale wolfberry

tomatoes. Various southwestern tribes have used the berries for food. In our area the species is confined to archeological sites. It is a "camp follower," brought in by previous inhabitants. Also called *tomatillo*. Found in pinyon-juniper woodland.

berry

PALE WOLFBERRY
Nightshade family SOLANACEAE
Lycium pallidum
Greek: *lykion*, a thorny shrub from Lycia, an ancient country in Asia Minor
Latin: *pallidus*, pale

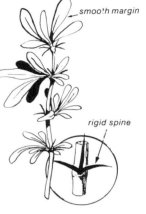
smooth margin

rigid spine

4. A few-branched shrub. Leaves with smooth to saw-toothed to holly-like margins. Spines at the nodes, 3- to 5-parted.

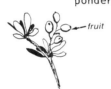
flower cluster

Erect, spiny shrub to 6 ft tall. Twigs reddish brown with 3- to 5-parted spines at nodes. Leaves alternate or clustered on the stem. Lower leaves become rigid spines. Leaves elliptical to long and narrow, margins smooth to spiny-toothed, holly-like. Flowers yellow. Fruit purple to red. Canyons of ponderosa pine and mixed conifer forests.

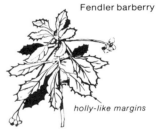
Fendler barberry

fruit

FENDLER BARBERRY
Barberry family BERBERIDACEAE
Berberis fendleri
Arabic: *berberys*, fruit of a barberry
Honors August Fendler (1813-1883)
German-born naturalist and explorer

holly-like margins

2. Leaves alternate or opposite on the stem.

5. Leaves palmately veined, 3- to 5-lobed. Spines usually in leaf axils.

CURRANT, GOOSEBERRY
Saxifrage family SAXIFRAGACEAE
Ribes spp.
Arabic: *ribas*, a type of sour plant

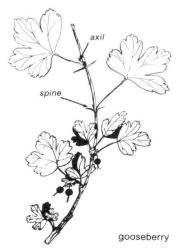
axil

spine

flower

Thicket-forming shrub. Leaves palmately veined and lobed, circular to kidney-shaped. Flowers tubular, pink to white. Fruits red. Fruits are edible. In canyons and on mountain slopes. Two species are found in the area:

WAX CURRANT *R. cereum* has no spines; foliage and flowers are somewhat sticky. (See also p. 41.)

GOOSEBERRY *R. inerme* has spines at the nodes.

gooseberry

5. Leaves not palmately veined.

   6. Branches with many spines, a low shrub. Leaves elliptical to oblong, underside hairy. Flowers white, small, fragrant, in dense clusters. (See also p. 35.)

BUCKBRUSH
Coffeeberry family RHAMNACEAE
Ceanothus fendleri

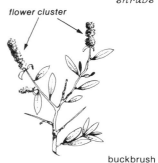

buckbrush

   6. Terminal thorns or axillary spines.

      7. Leafy twigs ending in sharp thorns. Leaves elliptical to oval. Margins saw-toothed.

      Shrub or tree to 6 ft tall, often thicket-forming. Petals 5, white to pink. Flowers 2 cm in diameter, 3 to 4 in umbrella-like clusters which appear before the leaves. Fruit reddish orange to red, somewhat globe-shaped. Along streams.

WILD PLUM
Rose family ROSACEAE
Prunus americana
Latin: *prunus*, plum tree

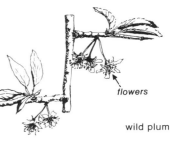

wild plum

      7. Spines in leaf axil, shiny, dark red.

      Large shrub or small tree growing to 15 ft. Members of the genus hybridize freely, so species are difficult to distinguish. Genus quite obvious because of shiny bark, zigzag twigs with dark red, nearly straight spines 5 cm long. Leaves oval, short-pointed, wedge-shaped at the base. Leafstalks (petioles) long, greenish to red. Flowers with 5 petals, white. Fruit small, apple-like, red at first, becoming black or brown. Spiny branches make favorite nesting places for birds. Fruits are eaten by various species of birds. Hawthorn is the state flower of Missouri. C. *erythropoda* most common in this area. (See also p. 25.) Found along streams in mixed conifer forest.

HAWTHORN
Rose family ROSACEAE
Crataegus erythropoda
Greek: *krataigos*, flowering thorn
erythros, red; *podos*, foot

hawthorn

1. Stems and twigs without spines or thorns.

    8. Leaves in bundles at the nodes.

        9. Leaves lance-shaped, grayish green, smooth. Flowers blooming before the leaves open.

            Straggly, irregularly shaped shrub growing to 10 ft, forming thickets. Leaves to 2 cm, lance-shaped, edges slightly rolled under, usually opposite on the stem but sometimes in groups with young leaves at the base of the old. Flowers yellow, blooming before leaves appear. Male and female flowers on separate plants. Fruit bluish black, resembling a small olive. Hopi Indians used the stems for making digging sticks. Shrub is an important source of food for birds and deer. Found in lower canyons along streams.

            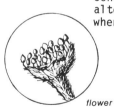
            female flower

            male flower

            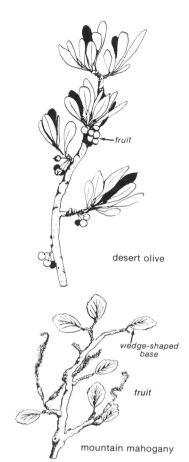
            fruit

            desert olive

            NEW MEXICO OLIVE, DESERT OLIVE
            Olive family OLEACEAE
            Forestiera neomexicana
            Honors Charles Forestier
            18th-century French physician

        9. Leaves oval, wedge-shaped at the base, with woolly hairs on the undersurface. Fruit tail-like.

            Erect, spreading shrub to 7 ft tall; most conspicuous when in fruit. Leaves oval, alternate, or in close bundles, particularly when first emerging. (See p. 43.)

            flower

            MOUNTAIN MAHOGANY
            Rose family ROSACEAE
            Cercocarpus montanus
            Greek: *kerkos*, tail; *karpos*, fruit
            Latin: *mons*, mountain

            wedge-shaped base

            fruit

            mountain mahogany

    8. Leaves alternate or opposite.

        10. Leaves opposite.

            11. Leaf margins toothed, bark shreddy.

                12. Leaf veining recessed, leaves soft, to 6 cm long.

                Erect shrub to 7 ft tall. Branches opposite, gray to reddish brown, bark shreddy. Leaves simple, opposite. Upper leaf surface bright green, underside downy white. Flowers white

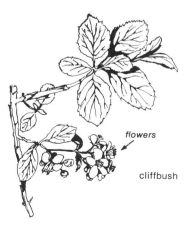
flowers

cliffbush

with 5 petals, fragrant. Well-drained and sunny sites, canyons.

CLIFFBUSH
Saxifrage family SAXIFRAGACEAE
Jamesia americana
Honors Edwin James (1797-1861)
American physician and explorer

shrubs

fruit

12. Leaves somewhat leathery, bright green, shiny, pale beneath. Flowers in pairs in leaf axils, pink, funnel-shaped.

    Shrub growing to 3 1/2 ft, in thickets. Leaves opposite, short-stalked, oval. Margins toothed or somewhat smooth. Flowers white to pink, nodding, in small clusters, funnel-like or tubular. Moist canyons and mountain slopes of mixed conifer and spruce-fir forests.

SNOWBERRY
Honeysuckle family CAPRIFOLIACEAE
Symphoricarpos oreophilus
Greek: *symphorein*, to yield together
karpos, fruit; *oros*, mountain; *philos*, loving

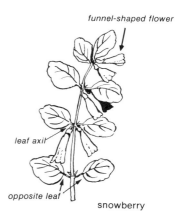

funnel-shaped flower

leaf axil

opposite leaf

snowberry

fruit

11. Leaf margins smooth, bark shreddy or smooth.

    13. Bark shreddy.

        14. Rigid shrub to 10 ft. Young branches white to straw-colored. Leaves without leafstalk (sessile).

            Shrubs with intricate branching, growing to 10 ft. Twigs reddish to straw-colored, becoming gray and shreddy. Leaves linear to oblong, margins rolled under. Petals 4, white, 1 to 2 cm. Dry, rocky slopes of White Rock Canyon and lower mesa canyons.

FENDLERBUSH
Saxifrage family SAXIFRAGACEAE
Fendlera rupicola
honors August Fendler (1813-1883)
German-born naturalist and explorer
Latin: *rupes*, rock; *colere*, to inhabit

flowers

opposite leaf

fendlerbush

flower

        14. Fine-textured shrub. Bark reddish, shiny.

            Delicately branched shrub to 4 ft tall. Stems red to chestnut-brown. Twigs of second season with conspicuous cross cracks. Leaves opposite, lance-shaped with smooth margins.

Upper surface green, smooth; lower paler. Flowers white, solitary on stem, 4 petals, fragrant. Stems were used for bows and arrows as well as pipe stems. Drained rocky soil of the canyons.

fruit

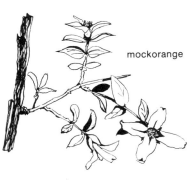
mockorange

MOCKORANGE
Saxifrage family SAXIFRAGACEAE
Philadelphus microphyllus
Greek: *philos*, loving; *adelphos*, brother
mikros, small; *phyllon*, leaf

13. Bark smooth. Leaves oval, 6 cm long. Flowers or berries in pairs in leaf axils, enclosed in leafy purplish bracts.

   Thicket-forming shrub, erect or spreading, to 10 ft tall. Twigs and branches paired. Bark yellow-gray. Leaves oval to oblong-oval, margins smooth. Leaf surfaces hairy with glandular dots, dark green above, pale below. Flower surrounded by oval-shaped bracts. Flowers yellow, tubular, 5-lobed. Berries dark purple or black. Moist canyons.

bracts
fruit

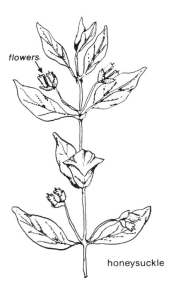
flowers
honeysuckle

BEARBERRY HONEYSUCKLE
Honeysuckle family CAPRIFOLIACEAE
Lonicera involucrata
Honors Adam Lonitzer (1528-1586)
German physician and herbalist
Latin: *involvere*, to wrap up

10. Leaves alternate.

    15. Leaves palmately veined.

       16. Plant a low undershrub. Leaves 5-parted, to 18 cm long. Flowers white, to 5 cm across.

          Low-growing shrub to 2 ft, occurring as single scattered plants or in dense patches. Stems smooth; old stems may be shreddy. Leaves to 18 cm long, palmately lobed with 3 to 5 lobes. Flowers in clusters of 2 to 9 at ends of branches. Fruit red, resembling large raspberry. Moist canyons.

fruit

thimbleberry

THIMBLEBERRY
Rose family ROSACEAE
Rubus parviflorus
Latin: *rubeus*, red
parvus, small; *florus*, flower

       16. Plant a moderate-sized shrub. Leaves not large, usually less than 5 cm long.

17. Leaves roundish, 3- to 5-lobed, star-shaped hairs on the veins and sepals. Bark shreddy.

Shrub to 4 ft tall. Stems slender, spreading, with bark shredding into strips. Leaves simple, oval to rounded, palmately 3 to 5 lobed, lobes toothed. Flowers in flat-topped clusters. Petals 5, white. Fruit yellow brown. Drained rocky soils of canyons.

NINEBARK
Rose family ROSACEAE
Physocarpus monogynus
Greek: *physa*, bellows; *karpos*, fruit
mono, one; *gyne*, woman

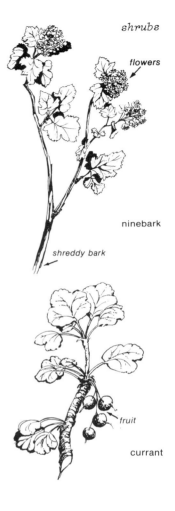

ninebark

shreddy bark

currant

17. Leaves roundish, 3-to 5-lobed, glandular hairs on leaves. Bark not shreddy.

Erect shrub to 6 ft. Leaves kidney-shaped or round. Petals white, tubular. Fruit red to black. Canyons and mountain slopes. (See also p. 36.)

WAX CURRANT
Saxifrage family SAXIFRAGACEAE
Ribes cereum
Arabic: *ribas*, a type of sour plant

15. Leaves not apparently palmately veined.

  18. Leaves less than 4 cm long.

   19. Leaves finely divided, gray-green. Flowers large, white. Fruit like a feather duster.

Scraggly, clump-forming shrub that blooms throughout the summer. Leaves wedge-shaped, pinnately divided into 3 to 7 blunt-tipped lobes. Flowers large, white, with 3 notches at tip of petal. Fruits feathery balls, reddish-tinged. Rocky to sandy soils along roadsides and in canyons.

APACHE PLUME
Rose family ROSACEAE
Fallugia paradoxa
Honors Virgilio Falugi, abbot and botanical writer
Greek: *para*, beyond; *doxon*, opinion
(referring to amazing similarity to rose flower)

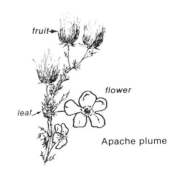

Apache plume

   19. Wedge-shaped, gray-green leaves with 3 teeth. Plant aromatic. Flowers tiny, in long sprays.

leaf

flower head
fruit

flowers
wedge-shaped leaf
big sagebrush
shreddy bark

Strong-smelling shrub, to 4 ft. Young stems hairy, older stems with shreddy bark. Very small flowers in heads arranged in long sprays. Shrub thrives best in alkaline-free soils. Early settlers soon learned to establish farms on sagebrush land rather than saltbush land. Wood produces an intense heat and pleasant aroma. It was an invaluable fuel to the Indians, particularly in arid areas without juniper or pinyon wood. The plant is a dominant shrub in many areas of the Great Basin and has been adopted by Nevada as its state emblem. The Spanish have used the leaves to cure rheumatism, croup, pains in the chest, common cold, indigestion, and headaches. Throughout pinyon-juniper woodland and sometimes into the ponderosa pine forest.

BIG SAGEBRUSH
Sunflower family COMPOSITAE
Artemisia tridentata
Honors Artemisia, botanist of ancient times
Latin: *tri*, three; *dens*, tooth

18. Mature leaves more than 4 cm long.

   20. Leaves wedge-shaped or egg-shaped.

      21. Flowers in a long plume; tiny, white. Remnants remaining on branches. Leaves oval, 5 cm long.

Shrub to 5 ft tall, multi-branched. Leaves egg-shaped, dark green above, woolly beneath, margins toothed. Flowers tiny, white, with 5 petals, in long plumes. Sepals densely hairy, turning pinkish with age, then russet, remaining on branches after petals and seeds have fallen. Seeds of this species have reportedly been eaten by southwestern Indians. Well-drained sunny sites in canyons and on mountain slopes.

flower

OCEAN-SPRAY
Rose family ROSACEAE
Holodiscus dumosus
Greek: *holos*, whole; *diskos*, disk
Latin: *dumosus*, bushy

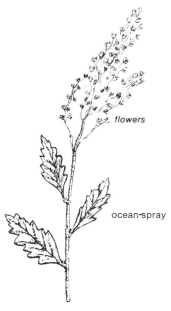

flowers
ocean-spray

      21. Plant with no plume of flowers remaining on branches.

         22. Plant with tail-like fruit. Leaves with a dense wool-like mat of hairs on the undersurface. Veins recessed.

shrubs

Erect, spreading shrub which grows to 7 ft tall; most conspicuous when in fruit. Fruit long-tailed, corkscrew-shaped. Twigs covered with waxy hairs which are lost as twig matures. Leaves oval-shaped, either alternate or in close bundles, margins toothed. Flower inconspicuous, greenish, 1 to 3 per leaf axil. Early inhabitants reportedly used the leafy twigs to keep bedbugs away. Its most extensive use is in making a dye. A soft reddish brown color is obtained from the root. Rocky areas of canyons and mesa tops. (See also p. 38.)

mountain mahogany

long-tailed fruit

MOUNTAIN MAHOGANY
Rose family ROSACEAE
Cercocarpus montanus
Greek: *kerkos*, tail; *karpos*, fruit
Latin: *mons*, mountain

22. Fruit fleshy, roundish, purple to blue-purple. Leaves dark green above, underside pale, thick and leathery.

Shrub to over 4 ft tall. Bark dark reddish brown to gray. Leaves oval, dark green. Flowers in clusters of 5 to 15. Petals white, showy, fragrant. Fruits spherical. Fruits are sweet and juicy, making good jellies and jams. Indians dried fruits for winter storage. Moist canyons.

fruit

SERVICEBERRY
Rose family ROSACEAE
Amelanchier utahensis
French: *amelanchier*, medlar-tree

serviceberry

20. Leaves longer than wide, not wedge-shaped.

   23. Leaves oblong in outline.

      24. Leaves somewhat leathery, margins lobed or wavy. Fruit an acorn.

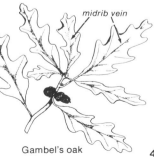

acorn

midrib vein

OAK
Beech family FAGACEAE
Quercus spp.
Latin: *quercus*, oak

Identification of oak species is made difficult by hybridization between them. Found in many habitats, including the pinyon-juniper woodland, ponderosa pine and mixed conifer forests. (See also p. 23.)

Gambel's oak

...ral species occur in the area:

GAMBEL OAK *Q. gambelii*. Lobed leaves with the lobes more than half way to the midrib vein. Leaves bright green above, sometimes velvety-hairy below. Ponderosa pine and mixed conifer forests. (See also p. 24.)

WAVYLEAF OAK *Q. undulata*. Smooth or wavy leaf margins cut no more than 1/4 the way to the midrib. Leaves blue-green with star-shaped hairs on the underside. Pinyon-juniper woodland, ponderosa pine forest.

GRAY OAK *Q. grisea*. Elliptical to oval gray-green leaves with star-shaped hairs on both surfaces. Leaves leathery with smooth or sharply toothed margins. Lower elevations.

wavy margins
wavyleaf oak

24. Leaves somewhat oblong to elliptical; edges very finely saw-toothed. Flowers white in dense clusters at ends of branches.

Thicket-forming shrub to 7 ft. Stems reddish brown to orange-brown. Leaves oval to oblong, toothed on margins. Leaf surface dark green, shiny above, pale green below. Flowers small, white, with 5 petals, arranged in cylindrical clusters. Fruit cherry-like, spherical, shiny, dark red-scarlet. Fruit astringent to taste but can be used for jams and jellies. Bark and leaves poisonous, particularly in the early spring. Inner bark yields a green dye. In canyons.

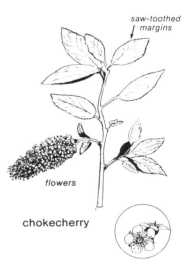
saw-toothed margins
flowers
chokecherry

CHOKECHERRY
Rose family ROSACEAE
Prunus virginiana var. *melanocarpa*
Latin: *prunus*, plum tree
Greek: *melas*, black; *karpos*, fruit

cherry-like fruit

23. Leaves long and narrow, linear to lance-shaped.

  25. Plant a subshrub, no taller than 2 ft.

    26. Leaves linear, smooth. Flowers in heads, yellow, showy.

Herbaceous plant, woody at base. Stems covered with a sticky substance. Leaves linear, 1 cm long. Flower heads arranged in flat-topped clusters. Each flower head with 4 or 5 disk flowers. Ray flowers 4 to 6, 2 to 5

flower heads
snakeweed

mm long. Throughout the pinyon-juniper woodland; indicator of overgrazing. (see p. 86.)

shrubs

snakeweed habit

SNAKEWEED
Sunflower family COMPOSITAE
Gutierrezia sarothrae
Honors Pedro Gutierrez, Spanish botanist
Greek: *sarothrae*, broom

26. Leaves linear or linear to lance-shaped, woolly-hairy. Flowers white to yellow, inconspicuous; arranged in axils at top of stems.

Shrub to 3 ft tall. Stems woody and branching at the base. Twigs silvery white, densely covered with woolly hairs. Flowers small, in dense clusters in the leaf axils, crowded along top of stem. Male flowers above the female flowers. Fruits densely silky-hairy. This plant is quite drought-resistant due to a deep taproot and lateral roots. It is considered an important winter forage because it is high in crude protein. Domestic animals, elk, mule deer, and rabbits feed on the plant. Found in dry, rocky canyons.

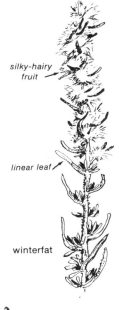

silky-hairy fruit

linear leaf

winterfat

WINTERFAT
Goosefoot family CHENOPODIACEAE
Eurotia lanata
Greek: *eurotia*, mold; Latin: *lana*, wool

25. Plant over 2 ft tall, not a subshrub.

27. Tall plant of moist locations. Leaves generally at least 2.5 cm long.

WILLOW
Willow family SALICACEAE
Salix spp.
Latin: *salix*, a willow

bud scale

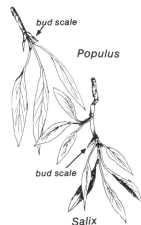

bud scale

Populus

bud scale

Salix

Identification of the different kinds of willows is difficult. Many hybridize and are similar in appearance. Mature leaves should be used in identification. The genus *Salix* is easily recognizable. It has only one bud scale forming a little cap over the bud; this is pushed off when the leaves emerge. Buds of other shrubs, particularly the closely related cottonwoods, *Populus* spp., have 2 or more scales which separate when bud growth begins. The plants of the willow family have inconspicuous flowers with no petals. These minute flowers are arranged in a drooping cluster called a *catkin* or *ament*, commonly called a pussy willow. Found in moist habitats.

Three common species occur in this area:

COYOTE WILLOW *S. exigua*. Leaves long and narrow, small saw-like teeth on margins. Twigs greenish, with short gray hairs. Scales of ament whitish green to pale green.

BLUE-STEM WILLOW *S. irrorata*. Leaves long, tapering at the end attached to stem, margins smooth. Ament turns black on drying.

BEBB WILLOW *S. bebbiana*. Leaves long, tapering at the end attached to stem, margins smooth, surfaces covered with hairs. Ament scales pale pink to rose-colored.

ament
willow

female flowers

27. Leaves gray-green, less than 1 mm wide. Dry habitats.

   28. Leaves linear.

      29. Fruits winged.

Erect, woody shrub to 5 ft tall. Gray-green leaves attached directly to stem. Branches covered with gray scale-like particles. Flowers light yellow; male and female flowers on different plants. Fruit with 4 winged bracts. Indians of the Southwest ground the seed to use as a baking powder in breadmaking. In the West stockmen consider this plant a good winter forage for cattle and sheep. It is a valuable winter browse for deer and antelope. Characteristically a shrub of alkaline or saline soils. Found in rocky areas of canyons and lower mesas.

fruit
female
four-wing saltbush

4-winged bract

FOUR-WING SALTBUSH
Goosefoot family CHENOPODIACEAE
Atriplex canescens
Greek: *atraphaxis*, Greek name for the orache, an edible weed; Latin: *canus*, white, hoary

   29. Fruits not winged.

      30. Bracts beneath flowers overlapping, in 4-6 vertical rows.

Shrub growing 3 to 8 ft. Bark shreddy, twigs with matted hairs. Leaves alternate, narrow, 2 to 8 cm long; margins smooth, surfaces densely hairy. Flowers in heads at ends of stems, yellow. Flower heads showy, but each

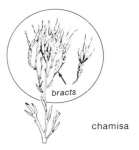
bracts
chamisa

individual flower small; no ray flowers present. This plant is often used to make a yellow vegetable dye for Navajo rugs. Pioneer along roadsides and in disturbed areas, growing in thickets. Pinyon-juniper woodland.

CHAMISA, RUBBER RABBIT BRUSH
Sunflower family COMPOSITAE
Chrysothamnus nauseosus
Greek: *chrysos*, gold; *thamnos*, thicket
nausea, seasickness

30. Bracts beneath the flowers equal in length.

Shrub growing to 1 to 2 ft. Stems hairy. Leaves 3 cm long, linear to lance-shaped. Head with 4 flowers, yellow. Heads in clusters at end of stems. Phyllaries 4 to 8 mm high. Dry canyons and mesas.

HORSEBRUSH
Sunflower family COMPOSITAE
Tetradymia canescens
Greek: *tetra*, four; *dymos*, together
Latin: *canus*, white or hoary

28. Leaves thread-like, or if 3-parted, segments threadlike.

Shrub to 3 ft. Leaves and stems covered with silvery hairs. Leaves thread-like, or if 3-parted the segments threadlike. Flower heads small, white to yellow; in loose clusters at tops of stems. In rocky canyons and in sandy soils of canyons at lower elevations.

SAND SAGEBRUSH
Sunflower family COMPOSITAE
Artemisia filifolia
Honors Artemisia, botanist of ancient times
Latin: *filum*, thread; *folium*, leaf

PARASITES AND SAPROPHYTES

This chapter describes parasitic and saprophytic species. *Parasites* have modified roots which enter the tissues of another plant and absorb nutrients. They weaken their host, often causing death. *Saprophytes* live on dead and decaying organic matter. Because these plants have no need to carry on photosynthesis, parasites and saprophytes usually have little or no green color.

1. Plant straw-colored, thread-like, parasitic on herbaceous plants.

   Twining herbaceous plants with yellow stems. Leaves reduced to minute scales. Flowers 5-lobed, yellowish to pink. Attached to various herbaceous plants. Found along roadsides. (See also p. 14.)

   DODDER
   Morning-glory family CONVOLVULACEAE
   Cuscuta umbellata
   Arabic: meaning unknown; Latin: *umbel*, parasol

1. Plant not thread-like.

   2. Plant attached to branches of trees.

      3. Plant attached to the branches of junipers. Stems yellow-brown.

         Parasites of junipers. Stems brittle. Leaves reduced to scales. Berries white to pink. Large quantities of these parasites attached to the tree may weaken or kill it. Birds relish the mistletoe berries and in the process of feeding transfer the sticky seeds from branch to branch and tree to tree, thereby spreading the parasite. Some Indian tribes used mistletoe medicinally. Others dried the berries and stored them for winter food. The berries of some species of mistletoe are poisonous.

         JUNIPER MISTLETOE
         Mistletoe family LORANTHACEAE
         Phoradendron juniperinum
         Greek: *phora*, thief; *dendron*, tree
         Latin: *iuvenis*, youth

      3. Plant attached to branches of ponderosa pine. Stems orange-brown, square in cross section.

juniper mistletoe

Parasitic on ponderosa pine. Flowers in clusters. Berries 2 to 3 mm wide. Heavy infestations of this parasite can kill trees.

parasites

female flowers

DWARF MISTLETOE
Mistletoe family LORANTHACEAE
Arceuthobium vaginatum
Greek: *arkeuthos*, juniper; *bios*, life
Latin: *vagina*, sheath

dwarf mistletoe

2. Plant not attached to the branches of trees.

    4. Flower 2-lipped, not symmetrical. Leaves scale-like.

        5. Flower parts in 3s. Side petals purplish brown.

            6. Lip (lower petal) white, 3-lobed, usually spotted.

Saprophyte, to 20 cm tall, purple to yellow. Sepals and lateral petals brown-purple, 3-veined. Lives in association with a fungus that decays stumps, logs, roots, and leaves. Because it obtains its food from the fungus it does not have green color. The underground stem is interwoven like a coral, giving the common name coralroot. Found in canyons, on mountain slopes.

SPOTTED CORALROOT
Orchid family ORCHIDACEAE
Corallorhiza maculata
Latin: *korallion*, coral; *rhiza*, root
macula, spot

spotted coralroot

            6. Lip (lower petal) purple with stripes. Margin entire.

habit

Saprophyte to 1 1/2 ft tall, purple to yellow. Sepals and lateral petals purple-brown, purple-veined. Lip yellow-brown to purplish brown with purple veins. Canyons and mountain slopes.

STRIPED CORALROOT
Orchid family ORCHIDACEAE
Corallorhiza striata
Latin: *striatus*, grooved

        5. Petals 5-lobed, snapdragon-shaped, purplish to pink. Leaves scale-like. Stems sticky.

striped coralroot

Perennial parasite to 15 cm tall. Stems purple to brown, yellow-brown, or pinkish, sticky. Leaves scale-like. Petals usually purple, to 2.5 cm long. Parasitic on species of buckwheat *Eriogonum* and sagebrush *Artemisia*. Commonly found under the BIG SAGEBRUSH *Artemisia tridentata* in the pinyon-juniper woodland.

CANCER ROOT
Broomrape family OROBANCHACEAE
Orobanche fasiculata
Greek: *orobanche*, a parasitic plant; *fascia*, bandage

4. Flower symmetrical. Leaves scale-like.

   7. Petals united. Flowers urn-shaped to spherical, hanging downward. Stems purple-brown, sticky.

Perennial saprophyte to 1 1/2 ft tall. Leaves scale-like, alternate. Flowers small, urn-shaped, petals 5. Found under pine trees in the ponderosa pine forest, occasionally in the mixed conifer forest.

PINEDROPS
Heath family ERICACEAE
Pterospora andromedea
Greek: *pteron*, feather, wing; *spora*, seed
Andromeda, in Greek legend, rescued from a sea monster by Perseus

   7. Petals not united. Stems cream-colored to reddish, turning black on drying. Flowers resembling an upside-down pipe.

Perennial saprophyte or parasite to 20 cm tall. Stems thick and fleshy. Leaves scale-like, reddish. Flowers nodding, either solitary or in clusters. Petals 3 to 5. Found in canyons and conifer forests.

PINESAP
Heath family ERICACEAE
Monotropa latisquama
Greek: *monos*, single; *tropos*, turn
Latin: *latis*, broad; *squama*, a scale

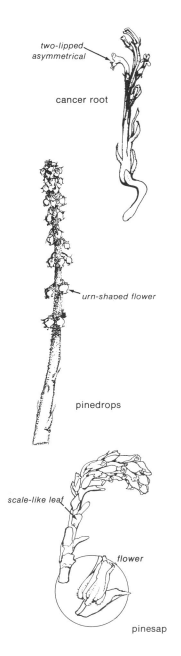

HORSETAILS

(Family EQUISETACEAE)

HORSETAILS *Equisetum* spp. belong to a single genus, and are the only surviving members of a once-large family. Other genera, now extinct, flourished during the Carboniferous, Devonian, and Triassic periods. Fossil evidence shows that many members of these genera attained tree size. Because all their relatives are known only as fossils, horsetails are called living fossils.

cone-shaped structure

HORSETAIL, SCOURING RUSH
Equisetum spp.
Latin: *equus*, horse; *seta*, bristle

Single-stemmed plant growing 1 to 3 ft. Stems green, jointed, hollow, ribbed or furrowed. Leaves scale-like, in a circle around the stem at each joint. Plants of two kinds, sterile and fertile. Sterile plants with branches at the nodes (these individuals being the "horsetails.") Fertile plants with a cone-like structure at tip, containing spores. The stems of horsetails contain large amounts of silica, giving the plants a rough texture. Early settlers called them scouring rushes, since the stems could be used to clean pots, pans, and floors. In marshes and wet soils throughout the area.

Species commonly found on the Pajarito Plateau include:

MEADOW HORSETAIL *E. arvense*, to 1 1/2 ft tall. Fertile plants unbranched, soon withering; sterile individuals with numerous lateral branches.

SMOOTH HORSETAIL, SUMMER SCOURING RUSH *E. laevigatum*, to 3 ft tall. Fertile and sterile plants similar, neither withering. Sterile plants with only a few lateral branches. A dark, narrow band apparent on the sheath at each joint.

SCOURING RUSH *E. hiemale*, to 3 ft tall. Fertile and sterile plants similar, neither withering. Mature stems with two ash-colored bands on the sheath at each joint.

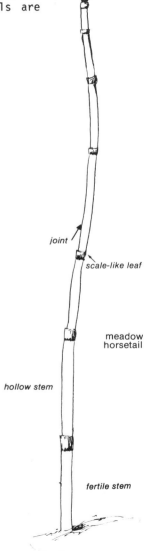

sterile stem

joint
scale-like leaf
meadow horsetail
hollow stem
fertile stem

GRASSES

(Family GRAMINEAE)

The true grasses constitute the largest family in the plant kingdom; their importance in nature, and to mankind, cannot be over-emphasized. Wheat, corn, oats, rye, barley, millet, sorghum, and sugar cane all belong to the grass family. Most domestic livestock and most of the major game animals of the world subsist largely on grass. Grasses can be distinguished from sedges, nutsedges, and rushes by the fact that stems of grasses possess swollen nodes. The stems of members of the other families are always uniformly smooth throughout. (See CATTAILS, RUSHES, AND SEDGES, p. 65.)

The flowers of grasses are highly modified. The parts include two lower bracts called *glumes* (corresponding to sepals) which hold one or more florets. Each floret is composed of upper bracts, *lemma* and *palea* (corresponding to petals), which in turn enclose a seed. Since the flowers are very small, usually less than 5 mm long, a magnifying lens is absolutely necessary for identifying grass species.

This chapter includes only a small fraction of the grasses of the Southwest woodlands. Most have such tiny florets that species identification is possible (at best) only with a microscope. The listed species have features distinctive enough to be readily recognizable.

The chapter is divided into three parts:

    I. PLANT CREEPING, SPREADING BY RUNNERS, p. 52
   II. PLANT UPRIGHT, SEEDHEAD UNBRANCHED, p. 53
  III. PLANT UPRIGHT, SEEDHEAD BRANCHED, p. 57

I. PLANT CREEPING, SPREADING BY RUNNERS.

1. Leaves in clusters at the nodes. Flowers hidden in leaf clusters.

    Mat-forming plant with wiry stems. Leaves stiff and sharp-pointed, to 5 cm long, clustered at nodes. Flowers hidden among leaves. Several florets in each flower. Lower bracts unequal; upper bracts with tufts of hairs on the margins. Blooms in summer and fall. Along roadsides and in disturbed areas.

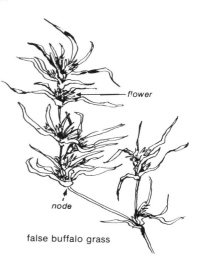

false buffalo grass

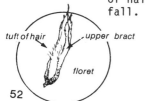

                    FALSE BUFFALO GRASS
                        Munroa squarrosa
        Honors William Munro (1818-1880)
                        British botanist
    Latin: *squarrosus*, stiff, spreading

1. Leaves not in clusters. Male flower heads extending above leaves.

>Sod-forming plant. Leaves gray-green, curly, hairy, rather soft. Male and female flowers on separate plants. Male flowers on one side of seedhead, which extends above leaves. Female flowers in bur-like clusters hidden in upper leaves. This species was the dominant grass of the Great Plains, and an important forage grass on the open range. The sod houses of early settlers were usually constructed of buffalo grass. Not native to the Southwestern woodlands, but introduced into lawns at the lower elevations.
>
>BUFFALO GRASS
>*Buchloe dactyloides*
>Greek: *boubalos*, buffalo; *chloe*, grass
>*daktylos*, finger; *oides*, like

grasses

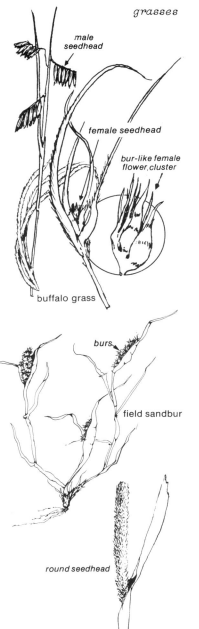

II. PLANT UPRIGHT, SEEDHEAD UNBRANCHED.

1. Flowers crowded on main stem.

    2. Flowers in stickery burs, often enclosed in the leaves.

        >Plant spreading, often growing in low mats, rooting at the nodes of the stems. Burs yellowish, about 1.5 cm long with stout spines. Usually found on sandy banks beside the Rio Grande.
        >
        >FIELD SANDBUR
        >*Cenchrus pauciflorus*
        >Greek: *kenchros*, millet
        >Latin: *paucus*, few; *flos*, flower

    2. Flowers not in burs.

        3. Seedhead round in cross-section. Not especially bristly.

            4. Flowers very small, less than 5 mm long, crowded the main stem.

                5. Flowers densely crowded on orderly seedhead to 10 cm long; rather like a miniature cattail in appearance. Plant to 3 1/2 ft tall.

                    >Flowers small, flattened, with 2 short awns. Hairs on the keels. Leaves flat, bright green. Stems swollen at base of plant. Robust plants often forming large clumps,

sometimes sown as a ground cover. Timothy is by far the most important hay plant in the U.S., exceeding even alfalfa in value. It is eaten by all grazing animals; even songbirds eat timothy seeds. Introduced from Eurasia. Fields and disturbed soil.

TIMOTHY
Phleum pratense
Greek: *phleos*, a kind of reed
Latin: *pratum*, meadow

5. Seedhead 6 cm long, somewhat shaggy, resembling an unkempt timothy. Stems with an abrupt bend.

Stems flattened with an abrupt bend; leaves flat, gray-green. Lower bracts dissimilar, 1 with a single central awn, 1 with 2 side awns. Upper bracts hairy, 1-awned. Rather spindly plants of rocky slopes and mesas.

WOLFTAIL, TEXAS TIMOTHY
Lycurus phleoides
Greek: *lukos*, wolf
phleoides, like timothy

4. Plant to 3 ft tall. Flowers 5 mm to 1 cm long, set in parallel rows at 45-degree angle to main stem.

Plants growing in stiffly upright clumps. Flowers on long stems above the leaves. Several florets per flower. Upper bracts tipped with a short awn, all green-striped. Introduced from Russia to reseed overgrazed ranges of the Great Plains. Resistant to cold and drought. Dry mesas.

DESERT OR RUSSIAN WHEATGRASS
Agropyron desertorum
Greek: *agras*, field; *puros*, wheat
Latin: *desertus*, abandoned

3. Seedhead cylindrical or flattened, very bristly at maturity.

6. Plant to 3 ft tall. Bristles attached to main stem rather than flowers.

Upright annual plant. Leaves broad with hairs on margins. Bristles of seedhead much longer than flowers. Flowers seed-like. Plant closely related to Italian millet. Seed makes

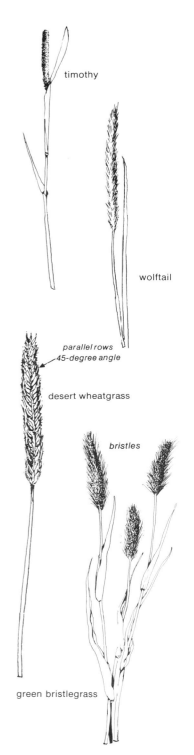

excellent food for game birds. Introduced from Europe. Roadsides and disturbed soil.

GREEN BRISTLEGRASS
Setaria viridis
Latin: *seta*, bristle; *viridis*, green

6. Awns attached to the flowers.

   7. Awns soft, whitish, about 1 cm long. Seedheads soft, silky to 15 cm long. Plant less than 2 ft tall.

      Flowers pale green when young, yellowish when mature. Lower bracts enclosing upper bracts. Introduced from Europe. Annual plants of the riverbanks.

      RABBITFOOT GRASS
      Polypogon monspeliensis
      Greek: *poly*, many; *pogon*, beard
      monspeliensis refers to
      Montpellier in southern France

   7. Awns long, 3 to 8 cm. Seedheads appearing very bristly at maturity.

      8. Plant less than 2 ft tall, rather stiffly upright.

         Flowers in pairs attached directly to main stem. Lower bracts reduced to long awns. Several florets in each flower. Seedhead falling apart at maturity. Blooms early spring through autumn. Widespread on mesas.

         BOTTLEBRUSH SQUIRRELTAIL
         Sitanion hystrix
         Greek: *sitos*, grain; *hystrix*, porcupine

      8. Plant about 5 ft tall. Seedhead bent.

         Flowers in groups of 3 or 4, attached directly to the main stem. Lower bracts narrow but not reduced to long awns. Awn at tip about 3 cm long. Seedhead intact at maturity, usually bent or slightly nodding. Seeds were used by Indians for food. Plant is susceptible to ergot, a poisonous and hallucinogenic fungus. Blooms late summer. Disturbed soil, usually roadsides.

         CANADA WILDRYE
         Elymus canadensis
         Greek: *elymos*, a kind of grain

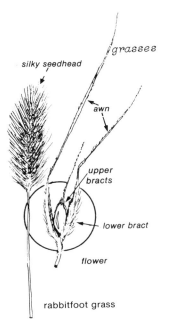

rabbitfoot grass

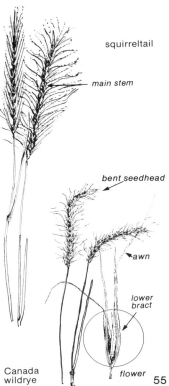

squirreltail

Canada wildrye

1. Flowers not crowded on main stem, usually overlapping, but main stem visible at a glance.

   9. Flowers in clusters attached directly to main stem, falling as a bundle.

      10. Flower clusters all on one side of main stem.

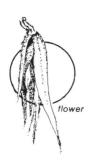

flower

Plant purplish when young, pale yellow in age. Flowers on one side of stem, often drooping; falling as a unit and leaving a short stub. An important forage grass. Indians bound the stems together to make a stiff whisk broom. Found in protected areas of pinyon-juniper woodland.

SIDE-OATS GRAMA
Bouteloua curtipendula
Honors Claudio (1774-1842) and
Esteban (1776-1813) Boutelou
Spanish horticulturalists
Latin: *curtus*, short; *pendere*, to hang

      10. Flower clusters evenly distributed around the stem.

Tufts of silky hairs at nodes of the stem and at base of the flowers. Leaves stiff, about 5 cm long. Short awns in flower cluster. Lower bracts broad and conspicuous. Plants of dry mesas.

GALLETA
Hilaria jamesii
Honors August St. Hilaire (1779-1853)
French, and Edwin James (1797-1861)
American botanists

   9. Flowers single, attached directly to main stem, only slight overlapping.

      11. Flowers set edgewise to main stem.

Only 1 lower bract on flowers. Introduced from Europe. This was the first grass to be cultivated as a distinct species in Europe. It is still an important forage species there. Meadows and waste places; often planted in lawns.

PERENNIAL RYEGRASS
Lolium perenne
Latin: *lolium*, darnel (a weed of grain fields)
per, through; *annus*, year

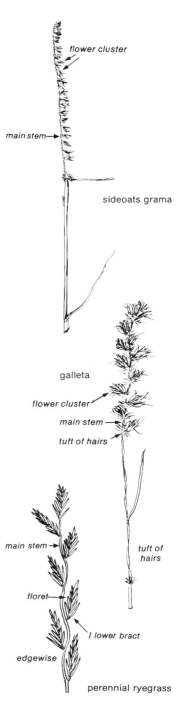

11. Flowers set against the main stem.

WHEATGRASS
Agropyron spp.
Greek: *agras*, field; *puros*, wheat

Seedhead long, often to 1 ft. Plant of open mesas. In addition to DESERT WHEATGRASS (p. 54), two additional wheatgrasses are common in the area:

WESTERN WHEATGRASS *A. smithii*. Plants spreading by underground stems (rhizomes.) Leaves upright, gray-green, with prominent ridges on their upper surfaces; waxy-looking.

SLENDER WHEATGRASS *A. trachycaulum*. Plant growing in large clumps. Leaves narrow, delicate-looking. One of five grasses seeded following the 1977 La Mesa Fire, it now covers the burned areas.

III. PLANT UPRIGHT, SEEDHEAD BRANCHED.

1. All side stems attached to main stem at the same point (digitate).

    2. Plant short, less than 2 ft tall. Side stems widely spreading.

        Leaves crowded at base. Flowers with short awns, often purplish, single along the side stem. Lower bracts broad, tapering sharply to awn. Disturbed soils, often in lawns.

WINDMILL GRASS
Chloris verticillata
Greek: *Chloris*, goddess of flowers
Latin: *vertere*, to turn

2. Plants tall, often to 7 ft. Side stems upright.

    Stems solid. Hairs on stems of seedhead. Flowers long, to 1 cm, reddish with tufts of shiny white hairs at base and a conspicuous bent awn. Seedhead breaking up at maturity. Big bluestem was the dominant plant of the tall grass prairies of the Midwest. Plant of the mesas, often found in burned areas.

BIG BLUESTEM, TURKEYFOOT
Andropogon gerardii
Greek: *andros*, man; *pogon*, beard
Honors John Gerard (1547-1612)
English botanist and surgeon

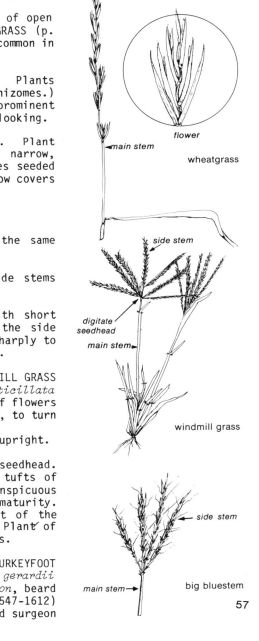

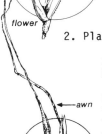

1. Side stems attached at different points along the main stem.

    3. Flowers all on one side of the stem. Seedheads resembling little flags or combs.

GRAMA GRASS
Bouteloua spp.
Honors Claudio (1774-1842) and
Esteban (1776-1813) Boutelou
Spanish horticulturalists

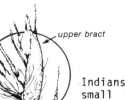

Indians of the Pajarito Plateau fashioned small brooms from these grasses to sweep metates. Zuni Indians strained goat's milk through the fine seedheads. Hopis wove the seedheads into their coiled baskets. Grama grass was also used in religious ceremonies; it was one of the plants tied to the wand for the squaw dance. Several species occur in the area:

BLUE GRAMA *B. gracilis*. Has 2 seedheads per stem; these curl at maturity in autumn. By far the most common grama grass; a dominant plant of the pinyon-juniper woodland.

HAIRY GRAMA *B. hirsuta*. Likewise has 2 seedheads per stem, and a spine extending beyond the flowers in a sharp point. Found in dry, exposed locations.

BLACK GRAMA *B. eriopoda*. Usually more than 2 seedheads per stem, and a tuft of white hairs where each seedhead is attached. Seedheads not curled when dry. Dry mesas.

(SIDEOATS GRAMA *B. curtipendula* is entirely different in appearance. See p. 56.)

3. Flowers disposed more or less equally on the stems. Not 1-sided.

    4. Seedhead very diffuse and wide-spreading. Flowers on long, slender stems.

        5. Plant about 2 ft tall. Flowers about 3 mm long, on long, slender stems at ends of dichotomous (paired) branches.

            Lower bracts membranous, shiny. Upper bracts appearing like a black seed covered with white hairs, terminating in a single awn. Seeds were used by Indians to make a mush. Mule

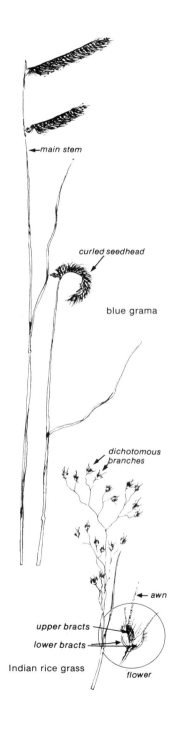

deer, mice, and chipmunks are known to relish the seeds, as do doves, quail, and green-tailed towhees. Dry mesas.

INDIAN RICEGRASS
Oryzopsis hymenoides
Greek: *oryza*, rice; *opsis*, resembling *hymen*, membrane; *-oides*, like

5. Plant to 3 ft tall. Flowers very tiny, less than 3 mm long, at the ends of fine side stems.

   Leaves encircling hairy stems. Seedhead much branched, often half the length of the plant; breaking off as a unit to become a tumbleweed. Plant annual, a conspicuous weed of roadsides and disturbed ground.

WITCHGRASS
Panicum capillare
Latin: *panicum*, millet; *capillus*, hair

4. Seedheads narrow or rather loose but not widely diffuse.

   6. Flowers dense on branches of seedhead.

      7. Flowers tiny, to 3 mm long. Stout, weedy, annual plant.

         Stems to 3 ft tall; often flattened in cross section, especially at the base. Some stems usually growing along the ground. Seedheads in tight clusters, bright green to purplish. Flowers small, about 3 mm long, spherical, with one side somewhat flattened; small bristles on bracts. Especially relished by waterfowl. Introduced from Europe. Disturbed ground; a common weed in gardens.

BARNYARD GRASS
Echinochloa crusgalli
Greek: *echinos*, hedgehog; *chloe*, grass
Latin: *crus*, leg; *gallus*, rooster

      7. Flowers to 1 cm long, awnless, in obvious clumps at the ends of side stems. Meadows and open mesas.

ORCHARD GRASS
Dactylis glomerata
Greek: *dactylis*, finger; *glomus*, ball

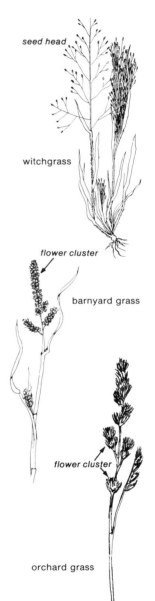

6. Flowers numerous or few, but not tightly packed on the side stems.

    8. Flowers long, narrow, and pointed, with a distinctive awn.

        9. Awns divided into 3 branches. Several species, of varied appearance, on open mesas.

                            THREE-AWN
                         Aristida spp.
                      Latin: *arista,* awn

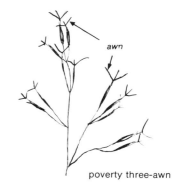
poverty three-awn

Mostly low grasses, distinguishable by length of awn branches and lower bracts. Occurring as isolated specimens on the open mesas of the ponderosa pine forest.

POVERTY THREE-AWN   *A. divaricata.* Small, spindly plant with flexible branches. Lower bracts are equal in length. Awns about 1 cm long. Usually grows in poor soil.

RED THREE-AWN   *A. longiseta.* Lower bracts unequal in length. Awns to 5 cm long.

ARIZONA THREE-AWN   *A. arizonica.* Seedhead narrow; lower bracts equal in length. Awns about 2 cm long.

    9. Awns very long, curling in age, often twisting around each other. Plant about 2 1/2 ft tall.

        Flowers very long and narrow with sharp point at base. Lower bracts longer than the flower. Awn twice-bent, twisted at base. Awns twist when wet, straighten as they dry, acting as an auger to drill the seeds into the ground. Rather rare but worth looking for. Open mesas to 7500 ft.

                            NEEDLE-GRASS
                            Stipa spp.
                    Greek: *stipa,* coarse flax
                 (alludes to flaxen appearance
                       of awns of some species)

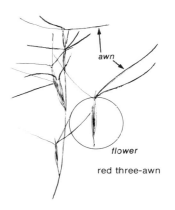
red three-awn

NEW MEXICO PORCUPINE GRASS   *S. neomexicana.* Has awns covered with feathery hairs.

NEEDLE-AND-THREAD GRASS   *S. comata.* Has smooth awns.

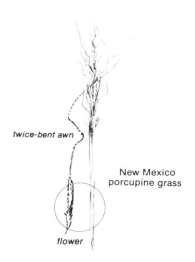
New Mexico porcupine grass

grasses

8. Flowers various, but not long, narrow, and hard.

   10. Flowers large, 1.5 to 4 cm long, with several florets in each flower.

      11. Lower bracts longer than florets.

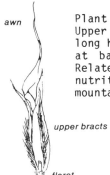

Plant to 3 ft tall, growing in large clumps. Upper bracts with 2 sharp points at top and long hairs at base. Awn flattened and twisted at base. Several florets in each flower. Related to cultivated oats *Avena sativa* and as nutritious to animals. Plants of high mountain meadows.

TIMBER OATGRASS
Danthonia intermedia
Honors Etienne Danthoione (d. 1815)
French botanist
intermedius, intermediate

      11. Lower bracts shorter than florets.

BROMEGRASS
Bromus spp.
Greek: *bromos*, oats

      12. Tall perennials, growing in sheltered canyons and meadows; common.

The brome grasses have the largest flowers of any of the grass family. They are excellent specimens for beginning the study of grasses. Leaves long and flat. Several florets in each flower; lower bracts unequal in length. Awns short or absent. Wild game animals--antelope, elk, moose, mountain goat, bighorn sheep, deer--relish the bromes.

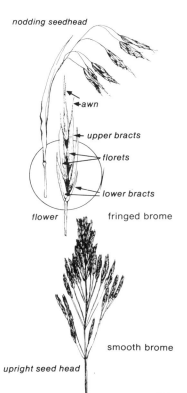

NODDING BROME *B. anomalus* and FRINGED BROME *B. ciliatus*. Have nodding seedheads. Canyons of the ponderosa pine and mixed conifer forests.

MOUNTAIN BROME *B. marginatus*. Has an upright flower and flattened seedhead. Canyons of the ponderosa pine and mixed conifer forests.

SMOOTH BROME *B. inermis*. Has a golden cast when mature, Flowers awnless on erect stems. Introduced from Europe; especially valuable as a soil-binder. Roadsides and disturbed areas of the ponderosa pine forest.

12. Weedy annual to 2 ft tall, growing in large dense masses. First grass to flower in the spring, usually dead by July.

   Seedhead drooping, often with purplish color. Leaves and flowers with soft hairs. Flowers nodding, to 2 cm long. Awns long, about 1.5 cm, spreading and stiff when old. One of the infamous "cheat" grasses: were an election held to name our worst weedy grass, downy chess would win hands-down. Introduced from Europe. In disturbed areas.

   DOWNY CHESS
   Bromus tectorum
   Greek: *bromos,* oats
   Latin: *tectorium,* roof
   (from sprouting of the seeds
   in thatched roofs in Europe)

10. Flowers small, less than 1 cm long.

    13. Seedhead loose, dense, green, tapering at top; very shiny at maturity. Plant about 2 ft tall.

        Stems hairy. Flowers small, awnless, less than 5 mm, shiny. Lower bracts as long as florets, with a green stripe down the center and transparent margins. Prominent on mesas of the ponderosa pine forest in early summer.

        JUNEGRASS
        Koeleria cristata
        Honors George Koeler (1765-1807)
        German botanist; Latin: *crista,* crest

    13. Seedheads narrow or open; any shininess due to hairs rather than flowers.

        14. Seedhead a loose pyramidal shape.

            Plant about 3 1/2 ft tall. Tuft of long white hairs where leaves are attached to stem. Flowers awnless, leaden-colored, only about 2 mm long but containing a loose black seed. Indians collected the seeds to make flour, which was used by the Mescalero Apaches to make bread and porridge, and by the Navajos to make dumplings, rolls, griddle cakes, and tortillas. Dry mesas.

            SAND DROPSEED
            Sporobolus cryptandrus
            Greek: *spora,* seed; *ballein,* to cast forth
            cryptos, hidden; *andros,* male

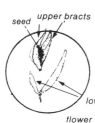
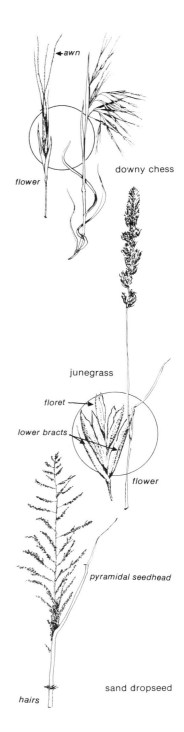

14. Seedhead not pyramidal.

    15. Seedhead oblong and loose.

        16. A tall, handsome plant growing to 5 ft; turning brown at maturity.

            Leaves flat. Flowers to 8 mm long, hairy, with a twisted, bent awn. A short, hairy side stem attached to the base of the flower. Occurs as solitary specimens on open mesas.

                          INDIAN GRASS
                          Sorghastrum nutans
                      Italian: *sorgo*, sorghum
            Latin: *astrum*, (a poor) imitation
                          nutare, to nod

        16. Plants to 2 1/2 ft tall. Seedhead grayish or lead-colored.

            Flowers tiny, about 3 mm long, on slender side stems. Seedhead grayish or lead-colored, elliptical in outline. Lower bracts gray, smooth. Upper bracts white, hairy on margins, giving flower a striped appearance. Common on mesas of ponderosa pine and mixed conifer forests.

                PINE DROPSEED, HAIRY DROPSEED
                       Blepharoneuron tricholepis
            Greek: *blepharon*, eyelid; *neuron*, nerve
            trichos, hair; *lepis*, husk or scale

    15. Seedhead narrow with short side stems.

        17. Flowers small, less than 3 mm, lacy, with fine awns.

            Plant to 3 ft tall. Branches lying close to main stem. One of the lower bracts three-pointed (visible with a strong lens), yellowish. Pollen sacs purplish in bloom, giving the flowers a mottled appearance. The common name is derived from the generic name and does not refer to mules. The dominant grass of the ponderosa forest.

                          MOUNTAIN MUHLY
                        Muhlenbergia montana
            Honors Gotthilf Muhlenberg (1753-1815)
                      German-born botanist
                    Latin: *mons*, mountain

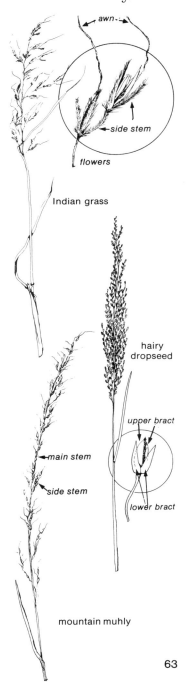

17. Flowers long, narrow, conspicuously silky-hairy, with a fine, bent awn.

   Flowers arranged in a zigzag manner on side stems. Two short hairy stems attached to base of flower. Hairs shiny white. Seedhead breaking between flowers at maturity. Flowering in late summer. Stems reddish in age. Together with big bluestem, this was the most important plant of the tall grass prairies of the Midwest. Bison, deer, antelope, burros, and domestic animals eat the young plants. Rodents and songbirds eat the seeds. Throughout pinyon-juniper woodland and ponderosa pine forest. Very common on mesas.

LITTLE BLUESTEM
Andropogon scoparius
Greek: *andros*, man; *pogon*, beard
Latin: *scopa*, broom

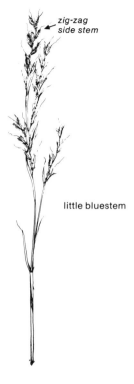

CATTAILS, RUSHES, AND SEDGES

CATTAILS, RUSHES, and SEDGES superficially resemble grasses, but differ in that their stems never display nodes. CATTAILS *Typha* spp. have solid, circular stems. RUSHES *Juncus* spp. have stems which are circular or flattened; SEDGES *Carex* spp. and NUTSEDGES *Cyperus* spp. have stems which are triangular in cross section. In members of the sedge family flowers are located at the axils of scales, which are grouped together to form a dense flower head (spike.) There are numerous species in three of the four genera, and any given one is therefore hard to identify. Only a few of the commonest are included here.

This chapter has three sections:

   I. STEMS CIRCULAR, TO 7 FT; FLOWERS IN DENSE CYLINDRICAL HEADS, p. 65
  II. STEMS CIRCULAR OR FLATTENED, WITHOUT NODES; FLOWERS IN LOOSE HEADS, USUALLY WITH TWO OBVIOUS BRACTS BENEATH THEM. p. 65
 III. STEMS TRIANGULAR IN CROSS SECTION; FLOWER HEADS FLATTENED OR ROUNDED, p. 66

I. STEMS CIRCULAR, TO 7 FT. FLOWERS IN DENSE CYLINDRICAL HEADS.

Perennial plants of marshy areas growing to 6 ft. Leaves alternate, very long, linear. Flowers in dense, tubular spikes at the end of the stem. Male flowers straw-colored, situated above the brown female flowers. CATTAILS are called the "outdoor pantry" because of their value as food for survival in the wild. Harrington's *Edible Native Plants of the Rocky Mountains* has suggestions on how to prepare roots, stems, and flowers for eating. In marshy areas, to 8000-ft elevations.

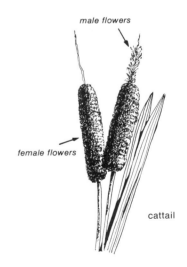

CATTAIL
Cattail family TYPHACEAE
Typha latifolia
Greek: *typhe*, cattail
Latin: *latus*, broad; *folium*, leaf

II. STEMS CIRCULAR OR FLATTENED, WITHOUT NODES. FLOWERS IN LOOSE HEADS, USUALLY WITH TWO OBVIOUS BRACTS BENEATH THEM.

RUSH
Rush family JUNCACEAE
Juncus spp.
Latin: *jungo*, join or bind

Fossils prove that this family has existed since Cretaceous time. The presence of rushes is a good indication of wet places. Flowers are brownish or yellowish, resembling dry, membranous scales. Rushes were and are used in basketry and mats. Along streams, near springs and seeps. Common local species include:

DRUMMOND RUSH *J. drummondii*. Stems to 1 1/2 ft tall, flowers 1 to 5, bracts to 3 mm, margins dark brown.

INLAND RUSH *J. interior*. Stems to 3 1/2 ft. Leaves 1/3 the length of the stem. Plant bears an inflorescence of many flowers; bracts 2, longer than the inflorescence.

TOAD RUSH *J. bufonius*. Stems to 20 cm, leaf blades 1 to 3 per stem. Inflorescence of single flowers, each flower with 2 bracts below it.

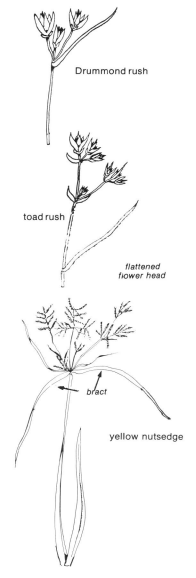

Drummond rush

toad rush

flattened flower head

III. STEMS TRIANGULAR IN CROSS-SECTION. FLOWER HEADS FLAT— TENED OR ROUNDED.

1. Spikes strongly flattened.

NUTSEDGE, FLATSEDGE
Sedge family CYPERACEAE
Cyperus spp.
Greek: *kypeiron*, marsh plant

Nutsedges have 3 ranks of leaves at the base of the stem. Two or more long leaf-like bracts lie at the base of the flower head, which is a long set of overlapping scales containing golden-brown flowers. The flowers are perfect, i.e., have both male and female parts. The roots have small edible tubers.

YELLOW NUTSEDGE *C. esculentus*. Stems to 3 ft. Bracts 3 to 6, longer than the spike. Moist ground.

yellow nutsedge

1. Spikes rounded.

SEDGE
Sedge family CYPERACEAE
Carex spp.
Latin: *carex*, sedge

Sedges *Carex* spp. have rounded flower heads atop triangular solid stems. Flowers are either male or female; frequently (though not

sedges

always) both male and female flowers occur in a common spike. Each female flower is borne in a small pouch called a *perigynium*. Stigmas are prominent and protrude from the long beak of the perigynium. Some sedges provide forage for both elk and domestic livestock. California Indians used sedge roots in basketry. In the local species cited below both male and female flowers occur in the flower spikes. (Some other species are dioecious, i.e., with male and female flowers borne on separate plants.)

MEADOW SEDGE *C. festivella*. Stems to 3 ft. Flower head egg-shaped, to 2.5 cm long, with inconspicuous bracts at its base. Perigynia (15 to 20) at top of spike, inconspicuous male flowers below them. Meadows within mixed conifer and spruce-fir forests.

FIELD SEDGE *C. praegracilis*. Stems to 2 1/2 ft. Male flowers at top of spike; perigynia (about 10) below them. Meadows and canyons at higher elevations.

perigynium

field sedge

CACTI

(Family CACTACEAE)

Cacti are succulent plants with large, fleshy stems but no apparent leaves. Through evolution the leaves have become spines; these are sometimes arranged in pits and sometimes located on raised areas called *areoles*. Some species have elongated or circular flattened stems called *joints*, others are spherical or cylindrical, with raised nipple-like areas called *tubercles*, or with parallel ridges called *ribs*. The family originated in the Americas, and cacti are considered the characteristic plant of American deserts, though they are not confined to deserts alone. Cacti found outside of the New World have been introduced by man. Cactus collecting for resale to fanciers has become a profitable business. As a result, many species are becoming rare and are considered endangered. The taxonomy of the Cactaceae is complex because many species readily hybridize. Only common species are discussed here.

1. Stems made up of flattened, long-cylindrical or club-shaped joints. Tiny spines in addition to long spines on areoles.

   2. Stems flattened.

      3. Joints 5 to 15 cm long.

         4. Fruits fleshy, purple. Spines strongly downward-slanting, usually on upper areoles only, 5 cm long. Flowers yellow.

habit

            Stems in clumps, to 15 cm tall. Joints oval to circular, 5 to 10 cm long. Spines white to grayish. Flowers yellow, often reddish at the base. Pinyon-juniper woodland, rocky canyon slopes.

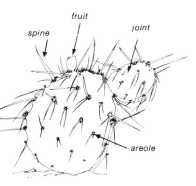

tuberous prickly pear

flower

            TUBEROUS PRICKLY PEAR
            Opuntia macrorhiza
            From Opus, a city in Greece
            where cacti are plentiful
            Greek: *makros*, long; *rhiza*, root

         4. Fruits dry, light brown, densely spiny. Spines in upper areole usually longer.

            Stems spreading, in clumps 15 cm tall. Joints circular to oval, to 10 cm long. Spines 6 to 10 per areole, downward-slanting. Flowers

cacti

yellow. Fruit dry, brown and very spiny. Pinyon-juniper woodland; dry, rocky canyons.

STARVATION CACTUS
Opuntia polyacantha
Greek: *poly*, many; *akantha*, spine

3. Joints 13 to 20 cm long.

   5. Fruit dry at maturity, spiny.

      6. Flowers yellow.

Stems clumped. Joints oblong-oval to somewhat circular. Spines on all areoles, whitish, curved upward or downward. Flowers yellow or pink. Fruit dry, brownish, densely spiny. Pinyon-juniper woodland.

fruit

HEDGEHOG PRICKLY PEAR
Opuntia erinacea
Latin: *erinaceus*, hedgehog

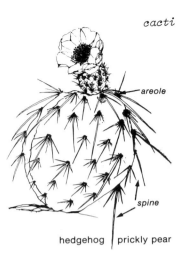
hedgehog | prickly pear

      6. Flowers red.

Closely related to STARVATION CACTUS, *O. polyacantha*; may not be a separate species. Areoles with 1 to 2 central spines. Fruit dry, spiny. Taxonomy of many species confusing and complex.

CLIFF PRICKLY PEAR
Opuntia rhodantha
Greek: *rhodon*, rose; *anthos*, flower

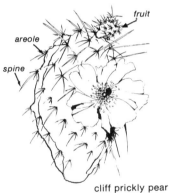
cliff prickly pear

   5. Fruit fleshy at maturity, red or purple, spineless.

Plant up to 3 ft tall. Stems sprawling. Joints egg-shaped, 1 to 15 cm long. Spines on most areoles, dark brown, to 5 cm long, 3 to 5 per areole. Flowers yellow, often marked with red. Fruit fleshy. Throughout the lower pinyon-juniper woodland.

fruit

PURPLE FRUIT PRICKLY PEAR
Opuntia phaeacantha
Greek: *phaios*, dusky
akantha, spine

2. Stems not flattened, stem long-cylindrical to club-shaped.

   7. Stems long-cylindrical, much-branched. Plants often to 6 ft tall.

purple fruit prickly pear

habit

Stems erect. Spines 10 to 30 per areole, red, pink or brown. Flowers to 7 cm wide, reddish purple. Fruit yellow, fleshy. This tree-like cactus is common on overgrazed ranges and other disturbed sites, such as archeological ruins. When the stem dies and the tissue decays, a hollow cylinder with a framework of diamond-shaped holes remains. These woody skeletons make interesting walking canes. Rocky or sandy soils, often in disturbed or overgrazed areas.

WALKING-STICK CHOLLA, CANE CHOLLA
Opuntia imbricata
Latin: *imbricatus*, covered with tiles

walking-stick cholla

7. Stems club-shaped. Plant matted, to 13 cm tall.

Spreading plant. Joints 5 cm long. Spines straight, 4-sided, gray, often tinged with pink. Flowers 5 cm across, yellow. Found on the sandy banks of the Rio Grande.

habit

CLUB CHOLLA
Opuntia clavata
Latin: *clava*, club

club cholla

1. Stems cylindrical or round, but not jointed.

    8. Stems with vertical or spiral ribs. Flowers on the rib next to a spiny areole. Fruit fleshy, spiny. Flowers at sides of the plant, below tip (apex). Flowers funnel-shaped.

        9. Flowers yellow-green, 2 to 3 cm long. Ribs 10 to 14. Stems seldom over 7 cm. Fruit green.

Stems oval. Spines reddish to brownish to white. Commonly found in rocky areas near the Rio Grande.

habit

GREEN-FLOWERED HEDGEHOG
Echinocereus viridiflorus
Greek: *echinos*, hedgehog
Latin: *cera*, wax; *viridis*, green
floris, flower

green-flowered hedgehog

        9. Flowers scarlet or purple, 5 to 9 cm long. Ribs seldom more than 12 to 13. Stem over 7 cm.

           10. Flowers scarlet or purple. Ribs 5 to 10 (usually 7). Stems often clustered in mounds.

cacti

habit   claret cup

fruit

Stems to 30 cm tall. Spines gray to tan, 3-angled, 2 to 2.5 cm long. Flowers red, to 7.5 cm long. Fruit red, fleshy. Common in rocky areas throughout the pinyon-juniper woodland.

CLARET-CUP HEDGEHOG
Echinocereus triglochidiatus
Greek: *treis*, three; *glochis*, arrowpoint

10. Flowers rose-purple to light purple. Central spine twice as long as lateral spines, often curved upward.

    Stems oval, to 25 cm tall. Ribs 8 to 10. Lateral spines white, gray, or yellow. Central spine circular in cross-section, curving upward. Flowers 7 cm long. Fruit green, becoming reddish on ripening. Pinyon-juniper woodland.

rib   Fendler's hedgehog

fruit

FENDLER'S HEDGEHOG
Echinocereus fendleri
Honors German-born explorer and
botanist August Fendler (1813-1883)

8. Stems without vertical or spiral ribs, tubercles not united. Flowers borne on tubercles, mostly near the top.

    Stems oval, 4 to 10 cm tall. Central spines red or brown at the tip, white at the base. Lateral spines white. Flowers pink to purple. Rocky or sandy soil of the pinyon-juniper woodland.

tubercle

PINCUSHION CACTUS
Coryphantha vivipara
Greek: *koryphe*, head; *anthos*, flower
Latin: *vivus*, alive; *parere*, to produce

pincushion cactus

FERNS

(Family POLYPODIACEAE)

Ferns are non-flowering plants reproducing by spores rather than seeds. These spores are borne in spore sacs or *sori* located on the undersides of fertile leaves. Each spore-case consists of a stalk and a small capsule; the structure responds to moisture. When moistened, the stalk bends back, then suddenly flips forward, hurling out the spores with considerable force. A spore which lands in a favorable spot germinates and grows into a small structure looking something like a tiny fallen leaf. This ultimately produces male and female cells. A fertilized female cell develops into an embryo, which then develops into another fern, thus completing the life cycle.

Ferns of different species have leaves ranging in size from a few millimeters to several feet. These leaves are called fronds. A *frond* consists of a stalk, called the *stipe*, and a *blade*. The blade is divided into segments called *pinnae*; each pinna resembles the leaflet of a compound leaf. The fronds of a growing fern emerge from underground in coiled form, and before uncoiling are called "fiddlenecks." Because of the nature of their life cycle, ferns are limited to moist habitats. They may be found along streams, in shady canyon areas, or beneath rocks.

I. PLANTS OF STREAMS AND MOIST CANYONS.

fiddleneck
habit

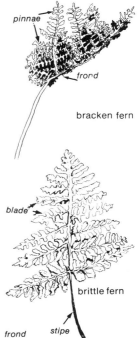
bracken fern

brittle fern

Fronds to 4 ft, triangular in shape, twice to three times pinnate. Sori only on the margins of the pinnae. An invader of overgrazed meadows. In canyons and on moist slopes of the ponderosa pine and mixed conifer forests.

WESTERN BRACKEN FERN
Pteridium aquilinum
Greek: *pteris*, fern (from *pteron*, wing); *aquila*, eagle

sori
spores
sorus
indusium

Grows to 2 ft tall. Fronds oblong to lanceolate, smooth, singly pinnate. Stipes very slender, brittle, with scales near the base which often fall off. Main stem of the frond straw-colored, stipe darker. Sori covered by a sheath, the indusium; not confined to margins of the pinnae. In moist shaded ledges, rocky slopes, and woods.

BRITTLE FERN
Cystopteris fragilis
Greek: *kystis*, pouch, *pteris*, fern
frangere, to break

May grow to 6 ft. Fronds 2- to 3-times pinnate; pinnae many. Stalks straw-colored, deeply grooved on 2 sides. Sori not confined to pinnae margins. Near streams, in canyons of the ponderosa pine and mixed conifer forests.

habit

LADY FERN
Athyrium filix-femina
Greek: *a-*, without; *thyreos*, shield
Latin: *filix*, fern; *femina*, woman

II. PLANTS AMONG ROCKS OF DRY CANYONS.

May grow to 1 ft. Fronds 1 to 4 times pinnate. Stipes brown to purple, wiry, sometimes zigzag. Sori on margins of the pinnae. Found under rocks in dry canyons at lower elevations.

CLIFFBRAKE
Pellaea spp.
Greek: *pellos*, dark (referring to stipe)

COMPOSITES

(Sunflower family COMPOSITAE)

The sunflower family, Compositae, is very large; its members share a unique flower structure. What appear to be single flowers are in fact flower heads composed of many small flowers, tightly packed. Individual flowers are usually of two kinds, *ray* and *disk* flowers. Ray flowers have five petals fused into a single petal that is strap-like; disk flowers are tubular, and end in five lobes. Instead of having flowers of both types, some species may possess only disk flowers, or only ray flowers. Individual flowers do not have sepals, but may or may not have a modified sepal called a *pappus*, which may take the form of scales or hairs. The flowers sit on a flat, conical, dome-shaped, or rounded structure called the *receptacle*. Papery scales separating the flowers form a *chaff* on the receptacles of some species. The entire flower head is encircled by a series of bracts called the *involucre*; a single bract is called a *phyllary*. The bracts are often of different sizes, and/or sometimes overlap. In many cases correct identification of a composite depends on observing the arrangement of the phyllaries.

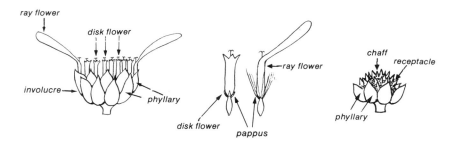

Because there are so many species and distinguishing between them many times depends on obscure characteristics, plants of this family are not easy for the novice to identify. To simplify the task, this chapter is divided as follows:

    I. GENERA WITH RAY FLOWERS ONLY, p. 75
   II. GENERA WITH DISK FLOWERS ONLY, p. 78
  III. GENERA WITH RAY AND DISK FLOWERS; RAY FLOWERS, YELLOW, ORANGE, OR RED, p. 84
  IV. GENERA WITH RAY AND DISK FLOWERS; FLOWERS WHITE, BLUE, PURPLE, OR PINK, p. 93

composites

I. GENERA WITH RAY FLOWERS ONLY.

In this group all the flowers in the head are ray flowers. The plants often exude milky juice when the stems or leaves are cut or crushed. This group is further divided into two sections according to the color of the flowers:

    A.   FLOWERS YELLOW, p. 75
    B.   FLOWERS WHITE, BLUE, OR PINK, p. 77

A. FLOWERS YELLOW.

ray flower

1. Leaves all at the base of the plant. Stems unbranched. One flower head per flower stem.

   2. Phyllaries overlapping in a graduated series.

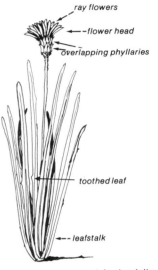
mountain dandelion

MOUNTAIN DANDELION
Agoseris spp.
Greek: *aix*, goat; *seris*, chicory

fruit

ORANGE-FLOWERED MOUNTAIN DANDELION *A. aurantiaca*. Stems to 2 ft. Leaves narrow, toothed to pinnatifid. Leafstalks purple. Flowers orange, turning purple with age. High mountain meadows.

PALE MOUNTAIN DANDELION *A. glauca*. Flowering stems to 1 1/2 ft, smooth or woolly near base of the plant. Leaves narrow, toothed to pinnatifid. Flowers yellow. Phyllaries sometimes spotted with purple. High mountain meadows.

   2. Phyllaries mostly in 2 series, outer shorter and directed downward. Stems hollow.

Perennial with leaves in a rosette at the base of the plant. Leaves lance-shaped, deeply pinnatifid, sparsely hairy to smooth. Ray flowers yellow, toothed at the tip. Fruit parachute-like. The dandelion is a good honey plant, has been cultivated for greens and used to make dandelion wine. The root has been used medicinally. Introduced from Europe, a pest in lawns. Common in meadows and along roadsides.

seedhead

COMMON DANDELION
Taraxacum officinale
Persian: *tarashgun*, chicory
Latin: *officina*, pharmacy

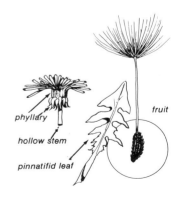
common dandelion

1. Leaves not all at the base of the plant; at least a few leaves or bracts on the stems.

   3. Flowers large, yellow. Seed head large and spherical.

   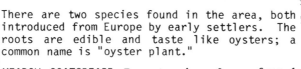

   SALSIFY, GOATSBEARD
   Tragopogon spp.
   Greek: *tragos*, goat; *pogon*, beard

   There are two species found in the area, both introduced from Europe by early settlers. The roots are edible and taste like oysters; a common name is "oyster plant."

   MEADOW GOATSBEARD *T. pratensis*. Grows from 1 to 2 ft tall. Leaves grass-like, twisted at the ends. Flowers yellow. Phyllaries in a single series of 8 or 9 bracts, shorter than the ray flowers.

   YELLOW SALSIFY *T. dubius*. Grows from 1 to 3 ft tall. Ray flowers lemon yellow. Phyllaries in a single series of 10 to 13 bracts, which are longer than the ray flowers.

   3. Flowers small, several clustered near the top of a leafy stem.

      4. Plants not over 1 ft tall. Stems stiffly hairy.

         Stems to 1 ft, stiffly hairy. Leaves basal, spatula-like, covered with stiff hairs. Leaves along the stem few, becoming smaller near the top. Flowers light yellow. Involucre cylindrical with 1 to 2 series of bracts. Commonly found in disturbed areas or along the banks of the Rio Grande.

         FENDLER'S HAWKWEED
         Hieracium fendleri
         Greek: *hierax*, hawk
         Honors August Fendler (1813-1883)
         German-born explorer and naturalist

      4. Plants growing up to 5 ft tall. Stems smooth.

         Annual with leaves unlobed to pinnatifid, spine-toothed, clasping the stem. Flowers yellow, in flat-topped, convex, or umbrella-like clusters at the top of a tall stem. Phyllaries overlapping. A naturalized weed found in disturbed soils.

         SPINY-LEAVED SOW-THISTLE
         Sonchus asper
         Greek: *sonchos*, sow-thistle, Latin: *asper*, rough

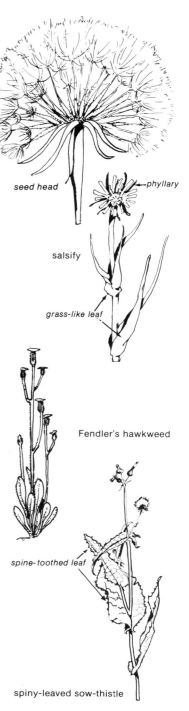

B. **FLOWERS WHITE, BLUE, OR PINK.**

composites

1. Flower heads showy (2.5 to 4 cm wide), blue, occasionally white.

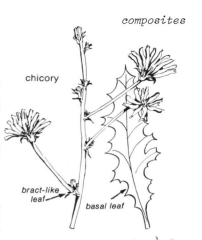

Perennial with branching stems growing to 3 ft. Leaves mostly basal, stem leaves becoming smaller and bract-like. Basal leaves pinnatifid, upper leaves clasping the stem, margins toothed. Flowers blue or occasionally white. Phyllaries in 2 series, outer series shorter and directed backward, with sticky hairs on the margins. Along roadsides.

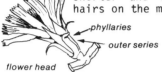

CHICORY
Cichorium intybus
Greek: *kichora*, chicory
Latin: *intybus*, chicory

1. Flower heads less than 2.5 cm wide, blue to pink.

   2. Plant up to 3 ft tall, leaves narrow to pinnatifid.

Stems to 3 ft tall, smooth. Lower leaves sharply pinnatifid; upper leaves with smooth to toothed margins. Flowers blue to blue-purple. Phyllaries overlapping into 3 or more series. Along roadsides, in gardens or other moist, disturbed soil.

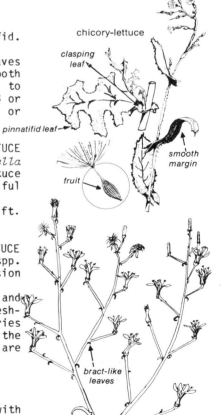

CHICORY—LETTUCE
Lactuca pulchella
Latin: *lactuca*, lettuce
pulcher, beautiful

   2. Plant with stiff, slender stems that grow to 2 ft. Leaves narrow, upper leaves bract-like.

WIRE-LETTUCE
Stephanomeria spp.
Greek: *stephein*, wreath; *meros*, division

Plants perennial, with stiff, rigid stems and alternate leaves. Flowers rose or flesh-colored. Phyllaries in 2 series, inner series longer than the outer. Commonly found in the pinyon-juniper woodland. Two species are found in the area:

S. tenuifolia. Stems hairy.

S. pauciflora. Stems smooth, sometimes with a tuft of woolly hairs at the base.

II. GENERA WITH DISK FLOWERS ONLY.

This group of composites has only disk flowers; all the flowers are tubular and symmetrical. This group is divided into 3 subgroups:

    A. PLANT OFTEN APPEARING SHRUB-LIKE, OR WOODY AT THE BASE, p. 78
    B. PLANT NOT SHRUB-LIKE; FLOWERS YELLOW, p. 81
    C. PLANT NOT SHRUB-LIKE; FLOWERS NOT YELLOW, p. 83

disk flower

A. PLANT OFTEN APPEARING SHRUB-LIKE, OR WOODY AT THE BASE.

1. Plant large. Flowers conspicuous.

    2. Flowers nodding, white to pinkish-purple. Leaves triangular to oval, conspicuously veined, resin-dotted.

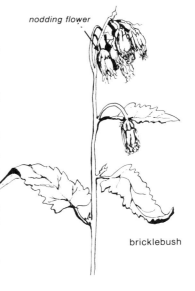
nodding flower

bricklebush

                    BRICKLEBUSH
                    Brickellia spp.
        Honors John Brickell (1749-1809)
              Irish-born naturalist

Stems to 3 ft, covered with short, stiff hairs. Canyons and mountain slopes. Species common in this area include:

CALIFORNIA BRICKLEBUSH *B. californica*. Heads of 6 to 20 flowers; greenish white, often tinged with purple.

flower head

B. grandiflora. Heads of 20 to 40 flowers, white tinged with green.

    2. Flower heads erect, not nodding. Leaves linear to triangular.

        3. Leaves large, taper-tailed.

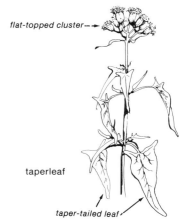
flat-topped cluster

taperleaf

taper-tailed leaf

Stems to 3 ft, strongly scented. Leaves triangular, having a taper-tail. Flowers pale yellow, heads in more or less flat-topped clusters. Phyllaries in 1 row, with thin, dry margins. Along roadsides, in old burn areas, and in canyons.

                    TAPERLEAF
                Pericome caudata
Greek: *peri*, around; *kome*, tuft of hair
                Latin: *cauda*, tail

phyllaries

flower head

3. Leaves linear to triangular but not taper-tailed.

   4. Leaves alternate, linear, and resin-dotted.

      Stems to 2 1/2 ft, finely hairy. Leaves linear to lance-shaped, prominently veined; margins toothed. Flowers white. Flower heads in flat-topped clusters. Phyllaries stiffly hairy and resin-dotted. Canyons and woods.

      FALSE BONESET
      Kuhnia chlorolepis
      Greek: *chloros*, green; *lepis*, scale

   4. Leaves opposite, oval-triangular.

      Stems to 2 1/2 ft, rather yellow-green in color. Leaves oval to triangular-oval. Leaf surface rough to the touch. Phyllaries in 2 rows, equal in length. Flowers 10 to 20 per head, white or faintly purple-tinged. Canyons and woods.

flower head

      DESERT THOROUGHWORT
      Eupatorium herbaceum
      Honors Eupator, king of Pontus
      Latin: *herba*, herb

1. Plant not large. Flowers small, to 5 mm wide, and inconspicuous. Foliage aromatic.

      WORMWOOD, SAGEBRUSH
      Artemisia spp.
      Honors Artemisia, botanist of ancient times

      Annual or perennial. Leaf margins smooth to variously lobed and dissected. Heads small, generally in loose clusters at the top of the stem. Phyllaries in 2 to 4 series, dry in texture, inner series thin, dry and membranous, or membranous on the margins.

      A number of species of *Artemisia* are found in the area. Most are woody at the base, aromatic, with tiny white, yellow, or brown flowers in loose clusters at the top of the stem. For the shrubby species see SHRUBS, pp. 42, 47. Plants treated here are generally herbaceous but woody at the base, or woody and low-growing.

✱ Woody species include:

BIGELOW SAGEBRUSH *A. bigelovii*. Grows to 1 1/2 ft. Leaves and stems covered with short silvery hairs. Leaves 3-toothed at the tip. Pinyon-juniper woodland.

ESTAFIATA *A. frigida*. Grows to 1 1/2 ft. Stems and leaves covered with short silky hairs. Leaves 3 times pinnatifid, making them appear feathery. An indicator of overgrazed or depleted ranges. Estafiata has been used medicinally to treat colds and as a diuretic. It is closely related to the wormwood used in Europe as source of medicinal oils and to flavor the liqueur absinthe. Widespread.

BIG SAGEBRUSH *A. tridentata*, see p. 42.

SAND SAGEBRUSH *A. filifolia*, see p. 47.

✱ Herbaceous species include:

WORMWOOD *A. ludoviciana*. Grows to 3 ft. Leaves silvery, margins smooth to deeply divided. The most common variety has shallowly lobed leaves. Found in canyons in the pinyon-juniper woodland and mixed conifer forest.

WORMWOOD *A. carruthii*. Grows to 2 ft. Leaves silvery, divided into filament-like sectors. Common in the pinyon-juniper woodland and in old fields.

RAGWEED SAGEBRUSH *A. franserioides*. Grows to 2 ft; stems grayish or tinged with red. Leaves bright green, dissected. Flowers on one side of the stem. Found under trees at higher elevations in the mixed conifer and spruce-fir forests.

FALSE TARRAGON *A. dracunculus*. Grows to 2 1/2 ft. Foliage without much odor. Stems reddish, smooth. Leaves linear to oblong lance-shaped; margins smooth to cleft. Found in abundance in old fields and abandoned sheep corrals. (See illustration p. 79)

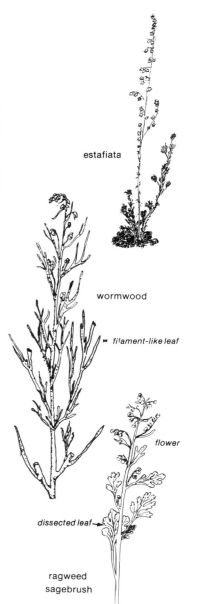

B. PLANT NOT SHRUB-LIKE; FLOWERS YELLOW.

1. Flowers tiny, in elongated clusters at the top of the plant. Plant weedy in appearance.

    2. Seedhead with hooked bristles. Leaves toothed or lobed.

        Weedy annual with stems to 2 1/2 ft, much-branched, rough and purple-dotted. Leaves widely oval, strongly wavy-margined or toothed, rough to the touch on both sides. Fruit 1 to 2 cm long, cylindrical, with prickles. Found in disturbed soils and sandy areas. Common along the banks of the Rio Grande.

        COCKLEBUR
        Xanthium strumarium
        Greek: *xanthos*, yellow
        Latin: *struma*, swelling

    2. Seedhead with spines or prickles. Female flowers small, male flowers above female flowers.

        3. Leaves lobed or dissected.

            Stems to 3 ft. Leaves alternate or opposite. Involucre composed of 5 to 12 partially united bracts. Female head 1-flowered and enclosed in a nut-like involucre having a single series of spines.

            RAGWEED
            Ambrosia coronopifolia
            Greek: *ambrosios*, immortal
            Greek: *korone*, crow; *pous*, foot
            Latin: *folium*, leaf

        3. Leaves toothed or pinnatifid.

            flower head

            Stems to 2 ft. Leaves usually alternate, toothed to pinnatifid. Involucres of 5 to 12 more or less united bracts. Fruit nut-like, with spines in several series.

            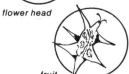
            fruit

            BURSAGE
            Franseria acanthicarpa
            Honors Antonio Franser
            Spanish physician and botanist
            Greek: *akanthos*, spine; *karpos*, fruit

composites

cocklebur

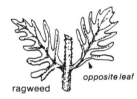
ragweed — opposite leaf

male flowers
female flowers
bursage
pinnatifid leaf

81

1. Flowers not small and generally solitary on the stem.

   4. Flowers, leaves, and stems sticky.

      Stems to 1 ft, straw-colored or tinged with reddish purple. Flowers yellow, sticky. Phyllaries in 5 to 6 rows, fleshy, sticky, and directed backward. Common along roadsides.

      GUMWEED
      Grindelia aphanactis
      Honors H. Grindel (1776-1836)
      Russian botanist
      Greek: *aphanes*, unseen; *aktis*, ray

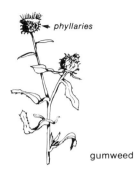
gumweed

   4. Flowers, stems, and leaves not sticky.

      5. Leaves pinnately divided.

         6. Leaves crowded near the base, 1 to three times pinnate. Herbage covered with loose, woolly hairs.

            Stems 1 to 2 ft. Leaves pinnate. Flowers yellow or creamy white. Phyllaries with yellowish tips. In the pinyon-juniper woodland.

            YELLOW CUT-LEAF
            Hymenopappus filifolius
            Greek: *hymen*, membrane; *pappus*, old man
            Latin: *filium*, thread; *folium*, leaf

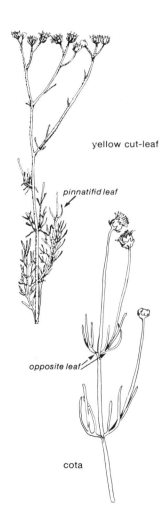
yellow cut-leaf

         6. Leaves along the stem pinnately parted or divided into linear segments. Stems and leaves smooth.

            Stems to 2 1/2 ft. Leaves once to twice divided into narrowly linear segments. Upper leaves smooth on the margins. Phyllaries in 2 series; lobes papery-margined. Flowers yellow, slightly nodding. Tea made from cota was the beverage of choice in the Southwest before commercial coffee and tea became available. When dried and then steeped in hot water, the plant makes an excellent iced tea; the beverage has been judged one of the best native herbal teas. The plant is also source of a yellow dye used to color native rugs. In some cases the tea acts as a mild diuretic. Pinyon-juniper woodland.

            COTA, NAVAJO TEA
            Thelesperma megapotamicum
            Greek: *thele*, nipple; *sperma*, seed
            megas, great; *potamos*, river

82

6. Leaves not pinnatifid.

> Stems to 3 1/2 ft. Leaves oblong-lance shaped; leafstalk winged. Upper leaves clasping the stem. Flowers yellow; heads 2 cm across, nodding. Found in mixed conifer, spruce-fir forests and in high meadows.
>
> BIGELOW GROUNDSEL
> *Senecio bigelovii*
> Latin: *senex*, an old man (referring to bald receptacle)
> Honors Dr. Jacob Bigelow, author of
> *The American Medical Botany*

C. PLANT NOT SHRUBBY; FLOWERS NOT YELLOW.

1. Leaves and phyllaries spine-tipped.

> THISTLE
> *Cirsium* spp.
> Greek: *kirsion*, thistle
>
> Biennial or perennial plants with spine-tipped or spine-toothed leaves. Heads are medium-sized to large, flowers pink, purple, or white. Corollas are tubular, deeply cleft. Species commonly found in the area include:
>
> PALE THISTLE *C. pallidum*. Stems and leaves covered with cobwebby hairs. Leaves smooth above and hairy below; lobes spine-tipped. Flowers greenish yellow. This thistle attracts beautiful orange butterflies. Commonly found in the high mountain meadows and along roadsides at the higher elevations.
>
> WAVYLEAF THISTLE *C. undulatum*. Stems to 5 ft. Stems and leaves covered with dense, wool-like hairs. Leaves spiny. Flowers rose-purple. Phyllaries taper to a spine, often twisted at the tip. Found along roadsides and in other disturbed areas.

1. Leaves and stems not spiny or spine-tipped.

    2. Flowers rose-purple, heads clustered along the stem.

    > Stems to 1 1/2 ft. Leaves linear, mostly at the base of the plant. Heads with 4 to 6 flowers. Phyllaries in several series,

phyllaries — disk flower

margins with hairs. Common, blooming in autumn in the pinyon-juniper woodland.

DOTTED GAYFEATHER
Liatris punctata
Liatris, etymology unknown
Latin: *punctum*, dot, point

2. Flowers white, in dense heads.

   3. Plant mat-like, with runners, less than 1 ft tall.

Perennials with oblong-oval leaves mostly at the base of the plant, covered with gray or white hairs. Flowers white. Phyllaries with a brown spot at the base and tip, white to pinkish. Meadows and open slopes of ponderosa pine and mixed conifer forests.

phyllaries — flower head

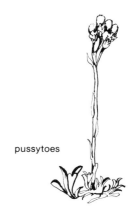
pussytoes

PUSSYTOES, CATSFOOT
Antennaria parvifolia
Greek: *anateinein*, to stretch forth
Latin: *parvus*, small; *folium*, leaf

   3. Plant over 1 ft tall, erect.

Perennials with densely matted, woolly hairs. Leaves linear to lance-shaped with hairs on the underside, smooth above. Flowers off-white; heads in flat-topped clusters. Open woods of ponderosa pine and mixed conifer forests.

flat-topped cluster

pearly-everlasting

flower head

PEARLY-EVERLASTING
Anaphalis margaritacea
Anaphalis said to be ancient Greek name for a similar plant
Greek: *margarites*, pearl

III. GENERA WITH RAY AND DISK FLOWERS; RAY FLOWERS YELLOW, ORANGE, OR RED.

These genera have both ray and disk flowers. Disk flowers may be the same color as the ray flowers, or may be brown, black, or reddish purple. The group is divided into two sections by the way the leaves are attached to the stem:

   A. PLANT WITH AT LEAST THE LOWER LEAVES OPPOSITE ON THE STEM, p. 85
   B. LEAVES EITHER ALL AT THE BASE OF THE PLANT OR ALTERNATE ON THE STEM, p. 86

A. PLANT WITH AT LEAST THE LOWER LEAVES OPPOSITE ON THE STEM.

1. Leaves divided into thread-like divisions, plant smooth.

    2. Ill-scented plant, leaves with dots, spine-tipped.

    Annual growing to 1 ft. Leaves once to twice pinnately divided, spine-tipped. Outer phyllaries green, inner phyllaries brown-green purple-tinged. Ray flowers small and inconspicuous, about 1 mm long. Found in disturbed soils of roadsides and trails.

    DOGWEED
    Dyssodia papposa
    Greek: *dysodes*, ill-smelling; *pappos*, old man

    2. Plant not ill-scented, leaves without dots.

    Perennial growing to 3 ft. Leaves twice pinnately divided, segments narrow to filament-like. Flowers bright yellow. Phyllaries in 2 series; outer series shorter than the inner, united at the base. Inner bracts papery at the margins. This species is often used to make a tea. It also makes a beautiful gold to gold-brown native dye. Throughout the pinyon-juniper woodland.

    GREENTHREAD
    Thelesperma trifidum
    Greek: *thele*, nipple; *sperma*, seed
    Latin: *trifidus*, thrice-divided

1. Leaves not divided, all opposite, or opposite at base but alternate above.

    3. Plant no more than 1 ft tall, lemon-scented.

    Annual, doubly branched and lemon-scented. Stems to 20 cm. Leaves long, narrow, and rather fleshy. Flowers yellow. Phyllaries in 2 series, with glandular dots. Found along roadsides and in sandy soil.

    FETID-MARIGOLD
    Pectis angustifolia
    Greek: *pectein*, to comb
    Latin: *angustus*, narrow; *folium*, leaf

composites

3. Plant coarse, usually over 1 ft tall.

    4. Heads numerous. Leaves with a leafstalk.

fruit

        Perennial with stems to 3 ft tall, branched, with fine hairs. Leaves lance-shaped to linear. Rays yellow, 1 to 2 cm, disk yellow. In ponderosa pine and mixed conifer forests.

GOLDEN-EYE
Viguiera multiflora
Honors L. G. A. Viguier, French botanist
Latin: *multus*, many; *flos*, flower

golden-eye

    4. Heads few, large and nodding. Leaves attached directly to the stem.

        Perennial with coarse stems, 1 1/2 to 5 ft, slightly hairy. Leaves at the base of the plant large, upper leaves leathery, about 4 pairs present, elliptical to lance-shaped. Phyllaries broadly oval to lance-shaped, blackening on drying. Rays yellow 2.5 to 3 cm long, disk flowers yellow. In open meadows at high elevations.

NODDING WOOD-SUNFLOWER
Helianthella quinquinervis
Greek: *helios*, sun; *anthos*, flower
Latin: *quinque*, five; *nervus*, nerve

B. LEAVES EITHER ALL AT THE BASE OF THE PLANT OR ALTERNATE ON THE STEM.

  1. Plant woody at the base (subshrubs).

    2. Leaves linear and narrow.

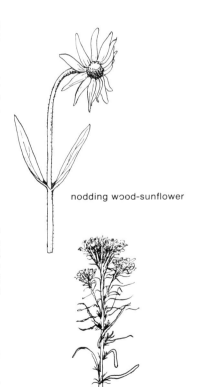
nodding wood-sunflower

        Perennial plant covered with a gummy exudate. Stems to 2 ft, leafy. Leaves alternate, linear. Flowers yellow. Heads in flat-topped clusters. Phyllaries overlapping, white-margined, tips green. This species is an aggressive invader where native vegetation has been removed by fire, overgrazing, or drought. Common throughout the pinyon-juniper woodland. (See also, p. 45.)

flower heads

SNAKEWEED
Gutierrezia sarothrae
Honors Pedro Gutierrez, Spanish botanist
Greek: *sarothrae*, broom

snakeweed

2. Leaves pinnately divided.

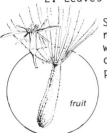
fruit

> Stems to 3 ft. Leaves pinnately divided into narrow sectors. Plant covered with dense, wool-like hairs. Flowers in flat-topped clusters. Pinyon-juniper woodland. (See also p. 89.)
>
> THREADLEAF GROUNDSEL
> *Senecio douglasii*
> Honors David Douglas (1798-1834)
> Scottish botanist and explorer

1. Plant not woody at the base.

   3. Disk flowers dark.

      4. Disk flowers dark red. Ray flowers yellow tinged with red.

         > Annual plant with stems to 2 ft tall, branched and covered with flattened hairs. Leaves oblong to lance-shaped, resin-dotted, slightly hairy. Ray flowers yellow at the tip and red at the base, notched at the tip. Blooms in masses along the roadsides of the pinyon-juniper woodland.
         >
         > FIREWHEEL
         > *Gaillardia pulchella*
         > Honors Gaillard de Marentonneau
         > French botanist
         > Latin: *pulcher*, beauty

fruit

      4. Disk flowers black or brown.

         5. Receptacle cylindrical or columnar.

            CONEFLOWER
            Ratibida spp.
            (meaning unknown)

            > Perennial plant with leafy stems, pinnately paired to divided alternate leaves. Ray flowers bright yellow to brown-purple. Phyllaries in 2 series, inner shorter than the outer. Two species are found in the area, particularly in disturbed or sandy soils:
            >
            > PRAIRIE CONEFLOWER *R. columnifera*. To 2 1/2 ft. Disk flowers on a columnar-shaped receptacle. Rays over 8 mm long. Leaves parted into 5 or more segments.
            >
            > CONEFLOWER *R. tagetes*. Stems to 1 1/2 ft covered with stiff hairs. Rays 4 to 6 mm long, directed downward.

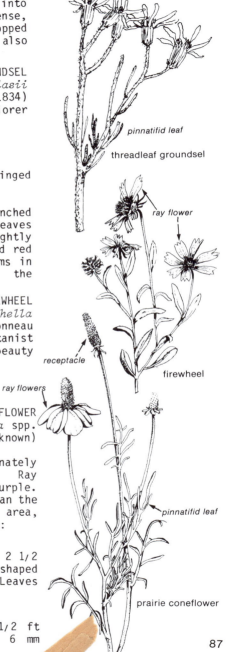

5. Receptacle cone-shaped or flat.

  6. Receptacle cone-shaped.

    7. Ray flowers yellow, directed downward, stems to 8 ft.

cutleaf coneflower

Perennial with large basal leaves divided into 3 to 7 oval to lance-shaped segments, smooth above, hairy below. Upper leaves 3-parted. Bracts oblong to oval, directed downward. Ray flowers yellow, 3 to 5 cm long, strongly directed downward.

CUTLEAF CONEFLOWER
Rudbeckia laciniata
Honors Olaus Rudbeck (1630-1702)
Swedish botanist
Latin: *lacinia*, flap, hem

    7. Ray flowers not directed backward. Leaves smooth on the margins.

Perennials with stems to 2 1/2 ft tall, hairy, and purple-dotted. Lower leaves oblong to lance-shaped, upper leaves linear. Ray flowers sometimes darker at the base, to 3 cm long. Found in moist canyons.

BLACK-EYED SUSAN
Rudbeckia hirta
Latin: *hirtus*, hair

  6. Receptacle flat.

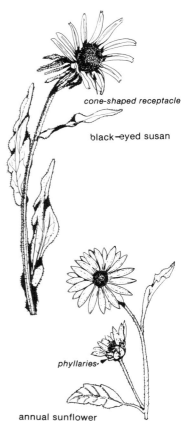

black-eyed susan

annual sunflower

SUNFLOWER
Helianthus spp.
Greek: *helios*, sun; *anthos*, flower

Annuals with leafy stems. Leaves mostly alternate. Phyllaries in several series. Sunflowers usually are found in disturbed soils of roadsides.

ANNUAL SUNFLOWER *H. annuus*. Grows to 9 ft tall. Stems hairy. Leaves oval to lance-shaped, coarsely toothed. Flower heads large, rays yellow, 2.5 to 5 cm long. Disk reddish purple, 2 to 5 cm wide.

PRAIRIE SUNFLOWER *H. petiolaris*. Annual, growing to 3 ft tall. Stems with hairs. Leaves oblong to lance-shaped, margins smooth to shallowly lobed. Rays yellow, to 2 cm long. Disks red-purple to brown-red.

3. Disk flowers yellow or orange. *composites*

    8. Individual flower heads in clusters.

        9. Phyllaries in 1 series, equal in length. Minute bracts beneath the heads.

GROUNDSEL
Senecio spp.
Latin: <i>senex</i>, old
(Makes reference to bald
receptacle)

Groundsels or senecios are best recognized by the flower heads, which have only a few ray flowers. Both disk and ray flowers are yellow. The various species are distinguished by leaf shape.

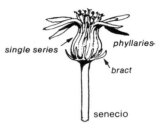

* Species appearing shrub-like, with leaves that are linear or have filament-like divisions, include:

THREADLEAF GROUNDSEL, *S. douglasii* var. *longilobus*. Grows to 3 ft tall; stems woody at base of plant. Leaves pinnately divided. Plant covered with dense wool-like covering of matted hairs. Flowers in flat-topped clusters. Involucres bell-shaped, with loose tufts of wool-like hair. Pinyon-juniper woodland. (See p. 87.)

S. multicapitatus. Grows to 3 ft, and is branched, with leafy tips. Hairs absent. Leaves irregularly pinnately divided into filament-like segments. Flower heads in flat-topped clusters; involucres without hairs. Pinyon-juniper woodland.

S. mullicapitatus

* Species that are herbaceous, with leaves entire to pinnatifid, include:

S. eremophilus var. *macdougalii*. Grows to 3 ft. Leaves lance-shaped, pinnatifid. Phyllaries black-tipped. Throughout the area.

FENDLER'S SENECIO or NOTCHLEAF BUTTERWEED *S. fendleri*. Grows to 1 3/4 ft. Stems covered with wool-like hairs in youth; these disappear with age. Leaves at base of the plant lance-shaped; margins smooth to pinnatifid. Upper leaves attached directly to the stem, linear to lance-shaped. Found in the mixed conifer forest.

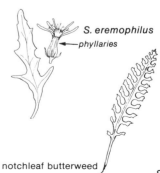

notchleaf butterweed

S. multiflorus. Grows to 1 1/2 ft. Stems leafy, smooth or with wool-like hairs only in axils of the leaves. Leaves mostly pinnatifid. Pinyon-juniper woodland.

NEW MEXICO BUTTERWEED *S. neomexicanus.* Grows to 1 3/4 ft. Stems leafy, covered with woolly, matted hairs. Leaves lance-shaped. Throughout the area.

9. Phyllaries in more than 1 series, unequal in length; mostly with white, papery margins and rarely with herbaceous tips.

New Mexico butterweed

GOLDENROD
Solidago spp.
Latin: *solidare*, to strengthen (in reference to supposed healing powers)

Solidago is recognizable by very small yellow flower heads arranged either in pyramidal clusters or on 1 side of the stem. There are numerous species; identification of individual species is often difficult.

WESTERN GOLDENROD *S. occidentalis.* Grows to 3 ft. Stems leafy, smooth, marked with fine longitudinal lines or furrows. Leaves with 3 resin-dotted nerves. Flower heads in more or less flat-topped clusters. Pinyon-juniper woodland and ponderosa pine forest.

ALPINE GOLDENROD *S. multiradiata.* Grows to 2 1/2 ft. Stems smooth to slightly hairy, often reddish brown. Leaves lance-shaped to oval, 1-nerved, with hairs on the margins and sometimes silky hairs at the base. Leaves at base of plant spatula-like, toothed on the margins. High meadows.

DWARF ALPINE GOLDENROD *S. spathulata.* Grows to 15 cm. Stems at base of plant lie on the ground, are often dark red. Leaves lance-shaped to oval, 1-nerved. High meadows.

FEW-FLOWERED GOLDENROD *S. sparsiflora.* To 2 ft. Leaves lance-shaped, 3-nerved, only slightly hairy or merely rough to the touch. Flower heads on one side of the stem. Pinyon-juniper woodland to mixed conifer forest.

alpine goldenrod

few-flowered goldenrod

8. Individual flower heads solitary, showy.   *composites*

   10. Most heads with 3 to 5 rays. Flowers becoming papery, leaves and stems woolly.

      Perennial growing to 20 cm, often woody at the base, covered with long silky or wool-like hairs. Lower leaves spatula-like or lance-shaped, covered with silky hairs. Upper leaves oblong to lance-shaped. Rays yellow, 3-lobed, becoming papery with age. Commonly found in sandy areas along the Rio Grande and pinyon-juniper woodland.

      paperflower

      flower head

      WOOLLY PAPERFLOWER
      Psilostrophe tagetina
      Greek: *psilos*, bare; *strophos*, twisted cord; *tagetes*, marigold

   10. Ray flowers more than 5.

      11. Erect plant.

         12. Leaves lyre-shaped.

            flower head

            Perennial growing to 1 ft tall. Leaves at the base of the plant, pinnately divided into 3 to 7 oval segments, margins toothed. Leafstalk as long as the blade. Upper leaves with 3 segments, margins smooth. Disk and ray flowers yellow. Phyllaries broad and oval. White Rock Canyon.

            LYRE LEAF, BERLANDIERA
            Berlandiera lyrata
            Honors J. L. Berlandier, Swiss botanist
            Greek: *lyra*, lyre

         12. Leaves not lyre-shaped.

            13. Leaves once, twice, thrice pinnatifid.

               Annual growing to 1 ft tall. Leaves finely divided in linear segments. Ray flowers yellow, to 2 cm long. Phyllaries in 2 to 3 series, with glandular hairs. Pinyon-juniper woodland to mixed conifer forest.

               WILD CHRYSANTHEMUM
               Bahia dissecta
               Honors Juan Francisco Bahi
               Spanish botanist
               Latin: *dissectus*, cut up

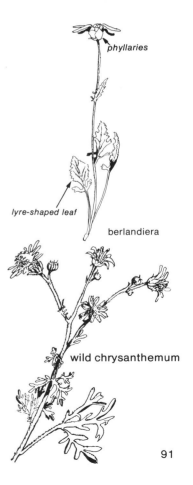
berlandiera

wild chrysanthemum

13. Leaves not pinnatifid.

    14. Plant with large showy flowers; leaves somewhat triangular.

        Annuals with stems to 2 ft, covered with matted, woolly hairs. Leaves oval to diamond-shaped, toothed; leafstalks winged. Ray and disk flowers yellow, to 2.5 cm, deeply cleft. Seeds flat and winged. Along roadsides and trails.

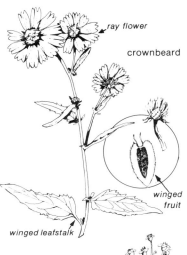

CROWNBEARD
Verbesina encelioides
Latin: *verbesina*, resembling verbena
From *Enkelados*, in Greek mythology a giant with a hundred arms

    14. Plant upright, ray flowers 3-toothed, often with orange veins. Foliage silvery.

ACTINEA
Hymenoxys spp.
Greek: *hymen*, membrane; *oxys*, sharp

Two species are commonly found in the area:

BITTERWEED *H. richardsonii*. Bushy perennial with a woody stem, growing to 1 1/2 ft. Leaves divided into 3 to 7 segments. Heads in flat-topped clusters.

PERKY SUE *H. acaulis*. Perennial with a leafless flower stalk. Leaves at the base, linear to lance-shaped, glandular-dotted, surfaces slightly to densely hairy. Flower heads solitary, to 2 cm broad.

H. argentea, also called PERKY SUE, differs from *H. acaulis* by having leaves along the stem.

11. Short, spreading plant, many-flowered.

    15. Leaves hairy, linear to spatula-like.

        Stems to 1 1/2 ft. Leaves linear to oval, with stiff hairs flat against the leaf. Ray and disk flowers yellow. Very common in pinyon-juniper woodland.

GOLDEN ASTER
Chrysopsis villosa
Greek: *chrysos*, gold; *opsis*, appearance
Latin: *villosus*, hairy

15. Leaves woolly, bristle-tipped.

composites

GOLDENWEED, GOLDEN ASTER
Haplopappus spp.
Greek: *haplos*, single; *pappus*, old man

Two species are found in the area:

SPINY GOLDENWEED *H. spinulosus*. Leaves oblong to spatula-shaped with teeth bristle-tipped. Phyllaries in 4 to 6 series, bristle-tipped.

H. gracilis. A definite green spot at the tip of each phyllary. Pinyon-juniper woodland.

bristle-tipped leaf

spiny goldenweed

IV. GENERA WITH RAY AND DISK FLOWERS: FLOWERS WHITE, BLUE, PURPLE, OR PINK.

1. Leaves finely dissected; flowers white.

    Perennial to 2 ft tall, covered with long, silky hairs. Leaves twice pinnately divided with segments very narrow, fern-like. Phyllaries with greenish line and brownish or blackish margins. Flower heads in flat-topped clusters. Common in canyons of ponderosa pine and mixed conifer forests.

    YARROW
    Achillea lanulosa
    For Achilles, who purportedly used
    the plant to treat the wounded
    Latin: *lana*, wool

flat-topped cluster

finely dissected leaf

yarrow

1. Leaves not finely dissected.

    2. Ray flowers very small, not longer than the involucre.

        3. Stem tall, up to 4 ft. Leaves alternate, linear.

            Annual plant with leafy stems. Off-white flowers in small heads arranged in a loose cluster at the top of a single stem. Phyllaries in 1 to 3 series. A common noxious weed of disturbed areas.

            HORSEWEED
            Conyza canadensis
            Greek: a kind of fleabane

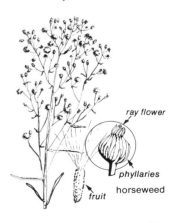

ray flower

phyllaries

fruit

horseweed

3. Stems up to 2 ft. Leaves opposite, upper sometimes alternate, twice to thrice pinnately divided.

>Annual plant to 2 1/2 ft, with smooth stems. Leaves twice to three times pinnately divided, segments no more than 2.5 mm wide. Flower heads to 9 mm high. Phyllaries in 2 series, inner shorter than outer. Ray flowers when present whitish, only 7 mm long. Seed long and narrow, with barbed horns. Dry canyons of the pinyon-juniper woodland.
>
>BEGGARTICKS
>*Bidens bigelovii*
>Latin: *bi*, two; *dens*, teeth
>Honors Dr. Jacob Bigelow, author
>of *The American Medical Botany*

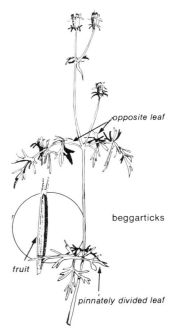

opposite leaf

beggarticks

fruit

pinnately divided leaf

2. Ray flowers large and showy, longer than the involucre.

4. Plant stemless.

>Perennial plant to 5 cm tall. Ray flowers white, disk yellow. Leaves linear to lance-shaped, nearly round in cross section because of inrolled margins, rough to slightly hairy. Phyllaries in 4 to 6 series with thin membranous margins. Very early spring bloomer in the pinyon-juniper woodland.
>
>EASTER DAISY
>*Townsendia exscapa*
>Honors David Townsend of Philadelphia
>botanical associate of Dr. William Darlington
>Latin: *ex*, out of; *scape*, stalk

Easter daisy

4. Plant with definite stems.

5. Ray flowers nearly as broad as long, leaves opposite.

6. Ray flowers rose-purple, disk yellow. Leaves dissected.

>Annual plant with stems to 2 1/2 ft, smooth. Leaves twice or thrice pinnatifid with filament-like divisions. Phyllaries in 2 dissimilar series, inner membranous on the margins, outer herbaceous. Seeds with stiff awns, which stick to clothing. Disturbed areas of roadsides and canyon bottoms.
>
>COSMOS
>*Cosmos parviflorus*
>Greek: *kosmos*, ornament
>Latin: *parvus*, small; *flos*, flower

phyllaries

fruit

cosmos

6. Ray flowers white, disk yellow. Leaf margins smooth to sinuately lobed.

> Perennial with stems to 1 ft, woody at the base, covered with gray or white hairs. Leaves narrow, margins smooth to sinuate-lobed, appearing grayish white. Outer phyllaries oval. Ray flowers to 13 mm, white with purple veins on the undersurface. Found at lower altitudes near the Rio Grande.
>
> PLAINS BLACKFOOT
> *Melampodium leucanthum*
> Greek: *melas*, black; *podos*, foot
> *leukos*, white; *anthos*, flower

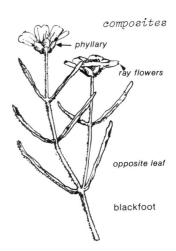

composites

blackfoot

5. Ray flowers definitely longer than wide.

   7. Phyllaries in 3 to 5 rows, graduated and overlapping, green near the tip, rays 8 to 35.

      8. Plants less than 15 cm tall, stems woody at the base. Leaves not more than 1 cm long.

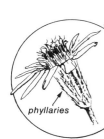

> Perennial with woody base. Ray flowers white, disk flowers yellow. Phyllaries green with papery margins, hairy near the tip. Stem leaves linear. Common in pinyon-juniper woodland.
>
> SAND ASTER
> *Leucelene ericoides*
> Greek: *leukos*, white; *erion*, wool
> *eidos*, form

sand aster

      8. Plants more than 15 cm tall. Leaves more than 1 cm long.

         9. Phyllaries broad, papery margined, with hairs.

> Perennial with stems no more than 15 cm tall. Leaves narrowly spatula-like to lance-shaped; to 5 cm long. Ray flowers violet or whitish, 7 to 12 mm long; disk flowers yellow. Flower heads solitary or in clusters of 2 to 3 at the ends of the stems. Phyllaries lance-shaped, silky-hairy with broad, white, papery margins appearing as though the edges had been torn. Pinyon-juniper woodland and ponderosa pine forest.
>
> TOWNSEND'S ASTER
> *Townsendia incana*
> Latin: *incanus*, hoary

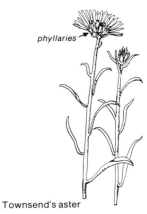

Townsend's aster

9. Phyllaries not broadly papery margined, either upright or directed backward.

   10. Phyllaries spreading or directed back at the tip. Usually papery at the base. Leaves with bristly teeth or lobes.

   ASTER
   Machaeranthera spp.
   Greek: *machaeros*, dagger; *anthos*, flower

   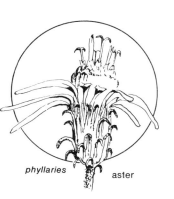
   phyllaries   aster

   Annuals or biennials with alternate leaves, much-branched. Lobes of leaves often bristle-tipped. Phyllaries leathery at the base and greenish at the tip, tips directed backwards. Ray flowers blue or purple, disk flowers yellow.

   Two species are common in the area.

   BIGELOW ASTER *M. bigelovii*. Stems to 3 ft tall covered with stiff hairs. Leaves oblong, lance-shaped, finely toothed and often glandular. Phyllaries in several series, linear, white at the base with green tips. Ray flowers purple. Along roadsides at lower elevations; mixed conifer forest.

   M. tanacetifolia. The only species with leaves once or twice pinnatifid. Ray flowers blue-purple to purple. Phyllaries parchment-like at base, tips green. Found in disturbed soils of the pinyon-juniper woodland.

   10. Phyllaries mostly upright to somewhat spreading. Leaves entire to toothed, but teeth not bristly.

   ASTER
   Aster spp.
   Greek: *aster*, star

   Bigelow aster

   Plants of the genus *Aster* are common; individual species are difficult to identify. They are generally fall bloomers, often growing profusely along roadsides or in other disturbed areas.

   ∗ Species of the area with white to pink ray flowers include:

   A. ericoides. Stems to 2 ft tall, rough hairy, branching above. Leaves linear to linear-oblong, with a bristle at each tip. Ray flowers white, sometimes pink, disk flowers yellow. Flower heads attached directly to and on the sides of the branches. Pinyon-juniper woodland.

composites

✱ Local species with blue or violet ray flowers include:

NEW ENGLAND ASTER *A. novae-angliae*. Stems covered with glands and with stiff hairs. Leaves clasping the stems. Ray flowers over 30, rose to violet-purple, disk flowers yellow. Phyllaries glandular, whitish near the tip. Common along roadsides throughout the area.

MARSH ASTER *A. hesperius*. Grows to 5 ft. Stems hairy. Leaves linear to lance-shaped. Ray flowers bluish, disk flowers yellow. Phyllaries wholly green, or sometimes green at the tip, white below. In moist canyons of the mixed conifer forest.

SMOOTH ASTER *A. laevis*. Grows to 3 ft. Stems hairy; leaves oval to oblong, disk flowers yellow. Phyllaries diamond-shaped, parchment-like at the base, green at the tip. Moist canyons and mixed conifer forest.

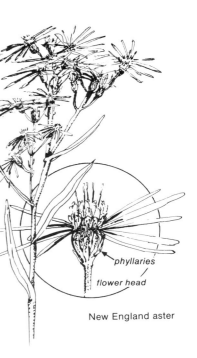

New England aster

7. Phyllaries in 1 to 2 rows, equal in length or nearly so, generally not overlapping, usually not obviously green at the tip. Ray flowers 25 to 150.

DAISY, FLEABANE
Erigeron spp.
Greek: *eri*, early; *geron*, old man

Erigerons, like the asters, are numerous and generally difficult to identify. They bloom throughout the summer, many in early spring.

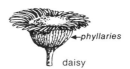

daisy

✱ Common species of the area with stems lying on the ground include:

TRAILING FLEABANE *E. flagellaris*. Biennial with prostrate stems which root at the tip. Plants covered with coarse, stiff hairs flattened on the stem. Leaves spatula-like to lance-shaped. Flower heads solitary, ray flowers white, to 1 cm long, disk flowers yellow. Phyllaries glandular. Common in disturbed soils of the pinyon-juniper woodland and ponderosa pine forest.

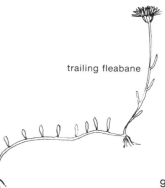

trailing fleabane

FLEABANE DAISY *E. divergens*. Biennial with branching stems to 1 1/2 ft covered with short, spreading hairs. Leaves lance-shaped, hairy. Ray flowers white, disk flowers yellow. Phyllaries sticky hairy. Found in pinyon-juniper woodland to mixed conifer forest.

✷ Species having erect stems include:

COMMON FLEABANE *E. philadelphicus*. Biennial growing to 2 ft. Stems covered with long, spreading hairs. Leaves lance-shaped, upper leaves clasping. Ray flowers white to rose-purple, to 2 cm; disk flowers yellow. Phyllaries smooth or with silky hairs, margins often purplish. Pinyon-juniper woodland to mixed conifer forest.

fleabane daisy

OREGON FLEABANE *E. speciosus* var. *macranthus*. Perennial, to 3 ft. Stems smooth to hairy. Leaves lance-shaped, hairy on the margins. Leafstalks of the lower leaves winged. Ray flowers blue, to 2 cm; disk flowers yellow. Phyllaries sticky-hairy. Ponderosa pine and mixed conifer forests, mountain meadows.

FLEABANE *E. subtrinervis*. Perennial growing to 3 ft. Stems covered with spreading hairs. Leaves lance-shaped, smooth on the margins, hairy on the veins. Ray flowers rose-purple to blue, to 2 cm long; disk flowers yellow. Phyllaries sticky-hairy. Ponderosa pine forest to alpine meadows.

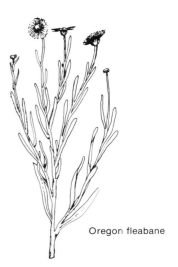

Oregon fleabane

SHOWY MONOCOTS

The subclass Monocotyledonae includes not only grasses, rushes, and sedges (treated in other chapters), but also a variety of species with much more showy flowers. The flower parts of these plants, usually called monocots, are in threes or multiples of three. Often the sepals and petals look alike (that is, the sepals are petal-like) and are referred to as a *perianth*. The leaves of monocots are parallel-veined. Many domesticated varieties of monocots such as the iris and many members of the lily family such as tulip, daffodil, and lily are favorites of gardeners. The orchids, considered by many the most beautiful of all, are also monocots.

The chapter is divided into three sections:

    I. FLOWERS WITH CONSPICUOUS PETALS OR SEPALS; SEPALS AND PETALS DIFFERENT. FLOWERS SYMMETRICAL, p. 99
   II. FLOWERS WITH CONSPICUOUS PETALS AND SEPALS; SEPALS PETAL-LIKE. FLOWERS SYMMETRICAL, p. 101
 III. FLOWERS WITH CONSPICUOUS PETALS AND SEPALS; FLOWERS ASYMMETRICAL, p. 104

I. FLOWERS WITH CONSPICUOUS PETALS OR SEPALS: SEPALS AND PETALS DIFFERENT. FLOWERS SYMMETRICAL.

1. Sepals drooping, petals erect. Stamens 3, encircled by petal-like styles. Leaves sword-like.

    Plants often growing in clumps. Flowers either solitary or in clusters, violet or blue streaked with purple. Petals erect; sepals petal-like, curved downward. Roots of this species are poisonous, containing an alkaloid called *irisin*. Although there are over 2300 different commercial varieties of iris, this is the only wild species in the area. Iris has long been identified with the *fleur-de-lis*, the symbol of the French monarchy. Found in open areas of the mixed conifer, spruce-fir forests, and in subalpine meadows.

    BLUE FLAG
    Iris family IRIDACEAE
    Iris missouriensis
    Greek: *iris*, rainbow

blue flag

1. Sepals and petals 3, erect; sepals not petal-like. Leaves various.

   2. A boat-shaped bract below clusters of flowers. Flowers blue.

      3. Leaves grass-like. Flowers blue, 7 to 10 mm across.

grass-like leaves

BLUE-EYED GRASS
Iris family IRIDACEAE
Sisyrinchium spp.
Greek: *sisyrinchion*, name used by Theophrastus for some plant

spathe / flower

Delicate little plant growing to 1 ft, usually shorter. Stems flattened and winged. Leaves narrow, grass-like. Flowers star-like, deep violet-blue with yellow center appearing like an eye. Leaf-like structure (*spathe*) below flowers. Species often intergrade. Two species commonly occurring in the area are:

S. demissum. Found near springs at lower elevations.

S. montanum. Found in high mountain meadows.

blue-eyed grass

   3. Leaves linear but not grass-like. Flowers 1 to 1.5 cm across.

      Stems to 1 1/2 ft. Leaves linear, smooth to finely hairy. Two sepals partially fused, somewhat colored. Petals blue, 2 lateral ones larger than the central one, narrowed at the base. A boat-shaped bract (*spathe*) 3 to 6 cm long beneath clusters of flowers. Flowers opening early in the day, then withering. Found in newly burned or otherwise disturbed areas in the ponderosa pine and mixed conifer forests.

spathe

DAYFLOWER
Dayflower family COMMELINACEAE
Commelina dianthifolia
Honors J. and K. Commelin
17th-century Dutch botanists
Greek: *di*, two; *anthos*, flower
Latin: *folium*, leaf

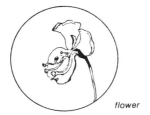

flower

dayflower

2. No boat-shaped bract beneath the flowers. Flowers white or purple.

monocots

SEGO LILY, MARIPOSA LILY
Lily family LILIACEAE
Calochortus spp.
Greek: *kalos*, beautiful; *chortos*, grass

Slender perennials with cup-shaped or tulip-shaped flowers, either white or purple. Leaves narrow, onion-like. Sepals green, petals colored, with yellow hairs or sometimes crescent-shaped spots at the base. Two species are found in the area.

SEGO LILY *C. nuttallii*. Flowers white. Found on dry mesas in the pinyon-juniper woodland in early spring.

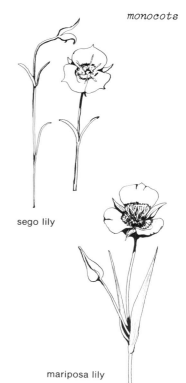
sego lily

crescent-shaped spot
flower

MARIPOSA LILY *C. gunnisonii*. Flowers purple. The bulb-like root (corm) of this species was a delicacy to western Indians. High in starch, it was frequently ground and made into bread. During the famine caused by drought, locusts, and frost (1848-1849), the plant was an important emergency food source to the Mormon pioneers who had followed Brigham Young to the Salt Lake Valley. For this reason the mariposa lily is honored as the state flower of Utah. Found in meadows at higher elevations in late summer.

mariposa lily

II. FLOWERS WITH CONSPICUOUS PETALS AND SEPALS; SEPALS PETAL-LIKE. FLOWERS SYMMETRICAL.

1. Leaves stiff, spine-tipped, in large rosettes at base of plant. Flowers on tall stalks which may or may not extend beyond the leaves. Flowers creamy white to greenish white. Fruits fleshy or dry.

YUCCA
Lily family LILIACEAE
Yucca spp.
West Indian: *yuca*, a yucca plant

rosette

Yucca was an extremely valuable plant to the Indians of the Southwest. The fibrous leaves were used to make baskets, mats, cloth, rope, and sandals. Flowers and fruits were eaten. The root, when pounded and soaked in water, provided a shampoo called *amole* which could be used for cleaning hair, feathers, or cloth. The suds were used to represent clouds in religious ceremonies. Two species are common in the area:

flowers

banana yucca

BANANA YUCCA *Y. baccata.* Leaves bluish green, 3 to 5 cm wide, to 2 1/2 ft long; tips often twisted. Margins with fibers curving backward. Flower stalk to 3 ft; flowers white, to 15 cm. Fruits fleshy, banana-shaped. Found on rocky slopes in canyons; not common.

NARROWLEAF YUCCA *Y. angustissima.* Leaves narrow, 4 to 8 mm wide, to 1 1/2 ft long; green. Fibers along margins. Flower stalk to 5 ft, flowers greenish white, to 6 cm long. Fruit dry. Common on mesas of pinyon-juniper woodland.

narrowleaf yucca

1. Leaves not stiff, or in rosettes at the base of the plant.

    2. Plant tall, 2 to 6 ft, coarse.

        Leaves large, oval, veins prominent. Flowers to 1.5 cm, greenish white to white, in large pyramidal clusters at the top of the stem. The roots of this and related species contain poisonous alkaloids acting as heart depressants, slowing the pulse. The plant has been known to poison livestock. The rootstocks are black, which is the origin of the name *Veratrum*. The powdered roots have been used as an insecticide. Found in moist meadows throughout the higher elevations. Also called CORNHUSK LILY.

false helleborine

flowers

                    SKUNK CABBAGE, FALSE HELLEBORINE
                          Lily family LILIACEAE
                            Veratrum californicum
Latin: *vere*, truly; *ater*, black

2. Plant less than 2 ft tall.

    3. Lower leaves alternate, upper leaves in a circle around the stem. Flowers large, orange-red.

        Stems to 2 ft. Leaves linear to lance-shaped, margins smooth. Flowers large, showy, red or orange-red with purplish black spots at the base. This strikingly beautiful plant is slowly being endangered by man's inconsiderate picking. It was once common, but is now confined to isolated areas. PLEASE DO NOT PICK.

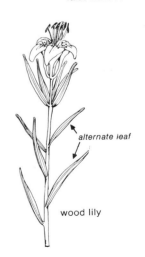

wood lily

                    ROCKY MOUNTAIN LILY, WOOD LILY
                          Lily family LILIACEAE
                               Lilium umbellatum
Latin: *lilium*, lily; *umbella*, parasol

1. Water-birch, *birch family*…p. 27.
Flowers of the birch family are hidden in catkin scales.

2. Willow, *willow family*…p. 45.
Pussy willows are also catkins with flowers hidden in the scales.

3. Poison ivy, *sumac family*…p. 33.
Shiny, beautiful, with white berries. POISONOUS! "Leaflets three, let it be."

4. Cancer root, *broomrape family*…p. 50.
Broomrapes live on dead or living plants so they don't need green leaves or stems.

5. Horsetail, *horsetail family*…p. 51.
Joint-stemmed relatives of horsetails dominated the earth 225 million years ago.

6. Stinging nettle, *nettle family*…p. 108.
Flowers of weedy plants often are tiny and inconspicuous. Nettle hairs STING!

7. Easter daisy, *composite family*...p. 94.
Each "petal" of a composite (sunflower or daisy family) is a separate flower.

8. Black-eyed Susan, *composite family*..p. 88.
Composite flowers are attached to a specialized central disk called a receptacle.

9. Dotted gayfeather, *composite family*...p. 83.
Not all composites look like sunflowers or daisies. some have distinctive flower heads.

10. Thistle, *composite family*...p. 83.
Not all composites have long, raylike flowers. Thistles have only short disk flowers.

11. Wild onion, *lily family*...p.103
Members of the lily family often have a bulb at the base of the stem, like an onion

12. Wood lily, *lily family*...p.102.
Lily family members have flower parts in sets of three, as here in two sets of three.

13. Blue-eyed grass, *iris family*…p.100.
Members of the iris family are closely related to grasses, lilies, and orchids.

14. Fairy slipper, *orchid family*…p. 105.
Flowers of the orchid family are highly nonsymmetric and exotically beautiful.

15. Butterflyweed, *milkweed family*…p. 122.
Flowers of the milkweed family look like little crowns.

16. Chimingbells…p. 124.
Flowers of the *forget-me-not family* are fused into little tubes, some long, some short.

17. Bluebell, *bellflower family*…p. 126.
Bellflower family members can have symmetric, bell shaped flowers…

18. Scarlet lobelia, *bell flower family*…p. 126.
…or wildly nonsymmetric, exotic flowers..

19. Rocky Mountain bee-plant...p. 127.
Flowers of the *caper family* have long stamens and long, string-beanlike fruit.

20. Wallflower, *mustard family*...p. 132.
All mustards have flowers with four petals in the shape of a cross.

21. Kinnikinnik, *heath family*...p. 29.
Heaths often have tiny flowers and fruit, wiry stems, and tough, smooth, shiny leaves.

22. Pipsissewa, *heath family*...p. 136.
Even heaths with showy flowers have the thick, smooth, shiny leaves.

23. Pinedrops, *heath family*...p. 50.
Pinedrops have tiny lantern-shaped flowers typical of many heaths.

24. Richardson's geranium...p.139.
Members of the *geranium family* all have fruit pointed like a heron's bill.

25. Deer's ears, *gentian family*…p. 138.
Plants of the gentian family have open, symmetric flowers…

26. Rose gentian, *gentian family*…p.138.
…or tubelike flowers. All have leaves opposite each other on the stem.

27. Scorpionweed, *waterleaf family*…p.141.
Members of the waterleaf family have flower heads coiled like a scorpion's tail.

28. Horsemint, *mint family*…p.142.
Mints have nonsymmetric flowers, opposite leaves, and a minty odor.

29. Lupine, *pea family*…p.146.
Members of the pea family have sweetpea-shaped flowers and beanlike fruit.

30. Western blue flax, *flax family*…p. 151.
Flaxes have flowers that fall off easily. Linen is made from the rigid stems.

31. Stickleaf, *blazing star family*…p.152. Blazing star flowers open for a brilliant display in the evening or during overcast weather.

32. Globe mallow, *mallow family*…p. 153. The long stamens of the mallow family fuse to form a tube.

33. Showy four-o'clock…p. 155. Beautiful flowers of the *four-o'clock family* have no petals, only colored sepals.

34. Sundrops, *evening-primrose*…p. 156. Four petals and T-shaped stamens are characteristics of the evening primrose family.

35. Skyrocket, *phlox family*…p.161. Phloxes have trumpet-shaped flowers; the length of the trumpet body varies greatly.

36. Shooting star, *primrose family*…p. 164. Despite some unusual flower shapes, all primroses have flower parts in fives.

37. Little red columbine, *buttercup*…p. 166.
Members of the buttercup family have varied, and often distinctive, flower shapes.

38. Pasque flower, *buttercup family*…p. 167.
Many flowers of the buttercup family have no petals. Colored sepals look like petals.

39. Rocky Mountain clematis…p.12.
This vining member of the *buttercup family* has no petals, only colored sepals.

40. Fendler meadowrue, *buttercup*…p. 168.
Some flowers of the buttercup family have bare stamens or pistils unprotected by petals.

41. Western baneberry, *buttercup*…p. 34.
"Avoid red or white berries," wise words concerning baneberries. POISONOUS!

42. Cow-parsnip, *carrot family*…p.182.
Flowers of the carrot family grow in umbrella-like clusters. Some members are poisonous.

43. James beardtongue, *figwort family*..p. 176. Wild snapdragons have the nonsymmetric flowers of the figwort family…

44. Scarlet bugler, *figwort family*…p.176. …as does the exquisite 'hummingbird' flower.

45. Indian paintbrush, *figwort family*...p.178. The beautiful red petals aren't petals at all, but colored bracts.

46. Valerian, *valerian family*…p. 184. This family represents a transition between the composites and "regular" flowers.

47. Pale wolfberry, *nightshade family*…p.35. Plants of the nightshade family have round, shiny berries. Most are poisonous.

48. Sacred datura, *nightshade family*...p. 179. POISONOUS! All parts of this nightshade are poisonous.

monocots

3. Leaves alternate or basal. Flowers white or pink.

    4. Leaves alternate.

SOLOMON'S PLUME
Smilacina spp.
Greek: *smilax*, bindweed (to which the plants bear no resemblance)

false solomon's seal (clasping leaf)

flower

Low, spreading plant. Leaves lance-shaped to elliptical. Flowers small, white. Two species are common in moist, shaded canyons:

FALSE SOLOMON'S SEAL *S. racemosa*. Stems to 2 ft. Leaves oval, clasping around the stem. Numerous white flowers, 2 mm long, at ends of the stems. Fruit to 6 mm in diameter, red with purple dots.

STAR FLOWER *S. stellata*. Stems to 2 ft. Leaves lance-shaped, not clasping the stem. Only a few white flowers, to 7 mm long, at ends of the stems. Fruit green with vertical blue stripes.

    4. Leaves basal.

        5. Flowers in umbrella-like clusters at the top of the stems.

WILD ONION
Lily family LILIACEAE
Allium spp.
Latin: *allium*, garlic

star flower

nodding onion (umbrella-like cluster, grass-like leaf)

Flowering stem to 2 ft tall. Leaves all grass-like, shorter than flowering stem. Flowers white to rose-colored; stamens protrude beyond petals. Wild onions were long used by native Americans for food and seasonings. On their expedition Lewis and Clark found them a welcome treat; the explorer Marquette depended upon them for survival. Wild species are related to cultivated onions and garlic. Leaves when crushed have the characteristic onion odor. Wild onions are prolific seeders, springing up in abundance in recently burned areas. Found throughout the ponderosa pine, mixed conifer, and spruce-fir forests. Two species found in the area are:

flower

NODDING ONION *A. cernuum*. Grows to 1 ft. Umbrella-like clusters of flowers nodding at top of the stem.

GEYER'S ONION *A. geyeri*. Grows to 1 ft. Flower clusters erect, not nodding.

5. Flowers along stem.

Stems growing to 2 ft. Leaves onion-like, bluish green. Flowers greenish white to yellow, sometimes tinged with purple; a long, leaf-like bract attached to the stem beneath each flower. This plant is poisonous to man and animals. Zygadenine, which the plant contains, is a powerful heart depressant; a dose of only 1/2 oz per 100 lbs of body weight is fatal. Indians and early settlers were sometimes poisoned by this plant when they mistook it for camas lily *Camassia*, wild onion *Allium*, or sego lily *Calochortus*.

DEATH CAMAS
Lily family LILIACEAE
Zygadenus elegans
Greek: *zygon*, yoke; *aden*, gland

III. FLOWERS WITH CONSPICUOUS PETALS AND SEPALS; FLOWERS ASYMMETRICAL.

The orchid family is the third or possibly second largest plant family in the world. Orchids may be as small as a pinhead or may grow to 1 1/2 ft tall. Some are saprophytic (living on dead and decaying organic material); others are parasitic (taking nourishment from other living plants). In the tropics many are epiphytes, living in trees high above the ground. Sepals and petals of the flowers are borne in threes. The uppermost sepal may differ from the other two by being larger, fleshier, and more richly colored and decorated. One of the three petals forms a distinctive structure called a *labellum*, or *lip*, often spotted or splashed with color. Orchids in this area are generally small and not brightly colored. Many are saprophytic.

1. Plants without green color.

CORALROOT ORCHID
Orchid family ORCHIDACEAE
Corallorhiza spp.
Greek: *korallion*, coral; *rhiza*, root

Single stem to 1 1/2 ft. Leaves small, scale-like. Flowers to 1 cm long, along upper 1/3 of stem. (See also PARASITES and SAPROPHYTES. p. 49.)

death camas

striped coralroot

monocots

Common species in this area include:

SPOTTED CORALROOT *C. maculata* with a white, spotted, 3-lobed lower lip.

STRIPED CORALROOT *C. striata* with a purple, striped, unlobed lower lip.

1. Plants with green color.

   2. Flowers reddish purple, slipper-shaped.

      Stems to 20 cm. Single leaf at the base of the plant. Flowers reddish purple, slipper-shaped. Lower petal large, inflated. Found in acid soils under pine trees; moist canyons and forests.

      FAIRY SLIPPER
      Orchid family ORCHIDACEAE
      Calypso bulbosa
      Greek: *Kalypsos*, a nymph of mythology
      bolbos, bulb

   2. Flowers white, green or purple; not slipper-shaped.

      3. Flowers greenish white with spur projecting downward from lip.

         Stems to 2 ft. Leaves 3 or more on stem. Flowers less than 1 cm, greenish white with spur projecting downward from lip. Found along stream banks or moist places in ponderosa pine and mixed conifer forests.

         BOG ORCHID, FRINGED ORCHID
         Orchid family ORCHIDACEAE
         Habenaria sparsiflora
         Latin: *habena*, thong; *spargere*, to scatter
         flos, flower

      3. Flower without a spur.

         4. Single leaf at base of stem.

            Stems to 15 cm. Flowers to 4 mm long, yellow-green; in tight clusters along flowering stem. Under pine trees in ponderosa pine and mixed conifer forests.

            ADDER'S MOUTH
            Orchid family ORCHIDACEAE
            Malaxis soulei
            Latin: *malaxare*, to soften
            soulei, origin unknown

fairy slipper

bog orchid

adder's mouth

4. Leaves 2 or more to a stem.

    5. Leaves basal.

        Stems to 20 cm. Leaves to 10 cm long, in a basal rosette, dark green with conspicuous white veins. Flowers white, to 5 mm long, along one side of the stem. Found in damp ponderosa pine and mixed conifer forests.

flower

        GIANT RATTLESNAKE PLANTAIN
        Orchid family ORCHIDACEAE
        Goodyera oblongifolia
        Honors John Goodyer (1592-1664)
        English botanist
Latin: *oblongus*, rather long; *folium*, leaf

    5. Leaves along the stem.

        Stems to 2 1/2 ft. Leaves to 15 cm, oval. Sepals greenish, petals greenish purple, lip marked with purple lines. Found near springs and streams. RARE. DO NOT PICK.

        HELLEBORINE
        Orchid family ORCHIDACEAE
        Epipactis gigantea
Greek: *epipeqnuo*, a plant which curdles milk
        Latin: *gigas*, giant

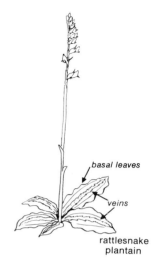
basal leaves
veins
rattlesnake plantain

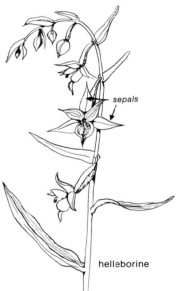
sepals
helleborine

WEEDS

What is a weed? It is a camp follower of agriculture and civilization. From the point of view of agriculture, weeds cause greater losses than do insects. They are usually prolific seeders and aggressive invaders of disturbed soils, but cannot withstand much competition from other plants. Weeds can be beautiful or dull. They may dress up roadsides or compete vigorously for garden space. A weed is any plant growing where it is not wanted; for example, a sunflower is a weed in the garden but a wildflower elsewhere. This chapter deals with generally unwanted lawn or garden occupants that have no showy flowers. Many do not appear to have flowers at all. ("Weeds" with showy flowers are treated in other chapters.) There are many species, only the most common are presented.

1. Plant creeping or lying on the ground.

    2. Leaves thick and fleshy.

A smooth, fleshy-stemmed, sprawling plant. Leaves alternate or opposite, thick and fleshy, spoon-shaped. Flowers yellow. This plant is called *verdolaga* by the Spanish-Americans and is used as a potherb or raw in salads. It is a hot-weather weed which spreads rapidly during the midsummer in rich soil of gardens and fields.

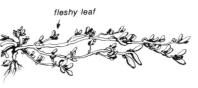
fleshy leaf

COMMON PURSLANE
Purslane family PORTULACACEAE
Portulaca oleracea
Latin: *portare*, to carry; *lac*, milk
olus, potherb

spoon-shaped leaf

common purslane

2. Leaves not thick and fleshy.

   3. Leaves opposite, oblong to oval, toothed above the middle. Stems exude milky juice when broken.

Mat-forming plant. Flowers tiny; sepals and petals absent. Leaves to 1 cm, narrow to oval, the bases usually unequal. Found in disturbed soil of roadsides and gardens.

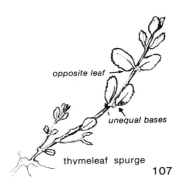
opposite leaf
unequal bases

THYMELEAF SPURGE
Spurge family EUPHORBIACEAE
Euphorbia serpyllifolia
Greek: after *Euphorbos*, physician to
the king of Mauretania
Latin: *serpyllum*, thyme-leaved

flower

thymeleaf spurge

3. Leaves alternate. Stems often purple, no milky juice when broken. Flowers in dense, chaffy clusters.

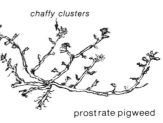
chaffy clusters

Prostrate annual. Stems much-branched, smooth or slightly hairy. Leaves oval, with crinkled edges. Bracts beneath the flowers spine-tipped. Found in gardens and along roadsides.

spine-tipped bracts

PROSTRATE PIGWEED
Amaranth family AMARANTHACEAE
Amaranthus graecizans
Greek: *amarantos*, unfading
Latin: *Graecia*, Greece

prostrate pigweed

1. Plant upright, not creeping or prostrate.

   4. Plant with STINGING HAIRS, growing in moist ground.

stinging hair

Erect plant growing to 6 ft tall. Leaves lance-shaped to oval, smooth on the upper surface, smooth to slightly hairy below, margins coarsely toothed. Flowers greenish, inconspicuous, in clusters near the top of the stem. The stiff hairs of this plant contain formic acid. When the tip of the hair penetrates the skin it produces a stinging sensation similar to an ant bite. However, it has been used as a potherb; cooking inactivates the formic acid. To prepare nettle soup see Harrington's *Edible Native Plants of the Rocky Mountains*. Found in patches along streams or in moist meadows.

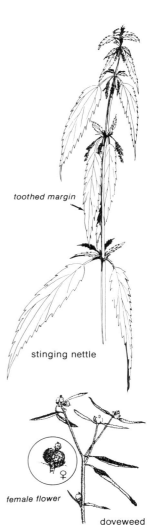
toothed margin

STINGING NETTLE
Nettle family URTICACEAE
Urtica gracilis
Latin: *urere*, to burn; *gracile*, slender

stinging nettle

   4. Plant without stinging hairs, generally on dry ground.

      5. Plant silvery gray with glandular star-shaped hairs. Strong-smelling.

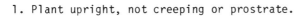

Erect annual, to 1 1/2 ft, with much-branched stems. Leaves long and narrow. Male and female flowers on separate plants. Sepals and petals 5, very small. Called *barbasco* by Spanish New Mexicans, who used it as an insecticide, particularly for bedbugs. Commonly found along roadsides and trails.

male flower

DOVEWEED, CROTON
Spurge family EUPHORBIACEAE
Croton texensis
Greek: *kroton*, a tick

female flower

doveweed

5. Plant smooth, or if hairy, hairs not star-shaped.   *weeds*

   6. Plant much-branched, forming dense clumps. Leaves bract-like, becoming spiny with age. Plant becoming a tumbleweed when dry.

      An intricately branched annual which may grow up to 6 ft tall in favorable conditions. Stems ridged, reddish. Flowers tiny, whitish, clustered at the base of the leaves. This weed is one of the most prolific in disturbed ground. It is a native of Russia and was brought into the United States in flax seed over 100 years ago.

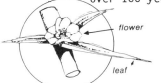

                              RUSSIAN THISTLE
                  Goosefoot family CHENOPODIACEAE
                                    Salsola kali
                              Latin: *salsus*, salty
                              Arabic: *al kahli*, alkali

bract-like leaf

Russian thistle

   6. Plant with definite leaves.

      7. Plant with large basal leaves, upper leaves smaller, alternate. Fruit winged.

                                    DOCK, SORREL
                         Buckwheat family POLYGONACEAE
                                    Rumex spp.
                              Latin: *rumex*, lance

arrowhead-shaped

         Several species of DOCK or SORREL are found in the area. The two most commonly encountered occur in disturbed ground or along streams.

         SHEEP SORREL *R. acetosella*. Plant to 1 ft tall. Leaves arrowhead-shaped. Flower heads reddish or yellowish, nodding.

         CURLYLEAF DOCK *R. crispus*. Plant to 3 ft tall. Leaves lance-shaped to oblong, large, with wavy margins. Flowers yellowish. Fruit triangular to roundish, heart-shaped.

sheep sorrel

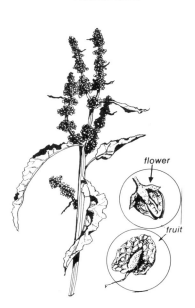

flower

fruit

curly leaf dock

      7. Plant without conspicuous basal leaves. Fruits various shapes but not winged.

         8. Leaves linear. Stems much-branched, often red. Mature seeds cottony.

Perennial, often with a woody base. Leaves narrow to lance-shaped, often with hairs on the margins. Flowers in dense leafy spikes. Introduced into the United States as an ornamental because of its bright red color in the fall. It has escaped cultivation and now grows as a weed along roadsides or trails and in gardens.

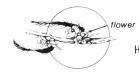
flower

SUMMER CYPRESS
Goosefoot family CHENOPODIACEAE
Kochia scoparia
Honors W. D. J. Koch, German botanist
Latin: *scopa*, broom

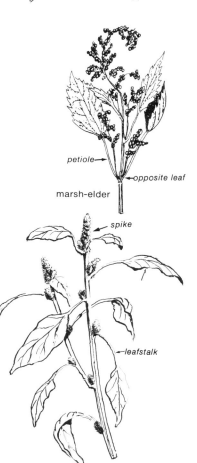

summer cypress

8. Leaves not narrow or if narrow then seeds not cottony.

   9. Lower leaves mostly opposite, gray-green, oval.

A very tall annual growing to 6 ft or more. Leaves coarsely toothed, large, to over 25 cm long. Leafstalk (petiole) long. Flowers inconspicuous. Contact with the plant may produce a dermatitis in some people. In moist ground along streams.

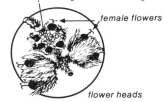
male flowers / female flowers / flower heads

MARSH-ELDER
Sunflower family COMPOSITAE
Iva xanthifolia
Iva, ancient name for
some medicinal plant
Greek: *xanthos*, yellow
Latin: *folium*, leaf

marsh-elder

   9. Leaves mostly alternate.

      10. Flowers in chaffy clusters. Stems smooth and red.

seed / fruit

Erect annual. Leaves dull green, egg-shaped to somewhat diamond-shaped, leafstalk long. Flowers green, small, in densely crowded spikes, 3 spiny bracts beneath each flower. Native of tropical America, where it has been cultivated for its seeds. One plant may produce as many as 100,000 seeds, making it a pest of gardens and other soils rich in nitrogen.

PIGWEED
Amaranth family AMARANTHACEAE
Amaranthus retroflexus
Greek: *amarantos*, unfading
Latin: *retro*, backward; *flexere*, to bend

pigweed

chaffy cluster

weeds

10. Flowers in tiny round clusters. Stems green.

   11. Plant smooth, covered with oily dots, odoriferous; green when fresh, bright red at maturity.

CHENOPODIUM
Goosefoot family CHENOPODIACEAE
Chenopodium graveolens
Greek: *chen*, goose; *podos*, foot
Latin: *gravis*, heavy; *olere*, to smell

   11. Plant covered with minute, mealy particles.

LAMB'S QUARTERS
Goosefoot family CHENOPODIACEAE
Chenopodium spp.
Greek: *chen*, goose; *podos*, foot

Of many species the following are frequently found in the area:

LAMB'S QUARTERS *C. album*. Leaves lance-shaped to egg-shaped; margins smooth. Flowers greenish, arranged in crowded clusters at the tips of stem branches. American Indians ground the small seeds into meal. The leaves make excellent salad greens when young; cooked, they taste much like spinach. In gardens and along roadsides.

FREMONT'S GOOSEFOOT *C. fremontii*. Differs from the preceding species in that the surface is less mealy, appearing bright green. Leaves are triangular to diamond-shaped.

chenopodium

lamb's quarters

HERBACEOUS (NON-WOODY) DICOTS

Plants belonging to the subclass Dicotyledonae are frequently called dicots. Dicots have flower parts (sepals and petals) in sets of four or five, or multiples thereof. Leaves are net-veined. If a plant has flower parts in fours or fives, it is a dicot, even though it may have narrow leaves that appear to be parallel-veined. (If, on the other hand, the flower parts are three or a multiple of three, consult the chapter on SHOWY MONOCOTS, p. 99.) Before continuing, you should also make sure that your plant definitely does not belong to another of the categories described in the preceding chapters. Woody dicots are keyed under VINES AND TRAILING PLANTS, p. 12, SHRUBS, p. 28, or TREES, p. 17. Herbaceous dicots without green color are in PARASITES AND SAPROPHYTES, p. 48. Those with ray, disk, or ray and disk flowers are in COMPOSITES, p. 74.

There are a number of families of herbaceous dicots. This key will be used to determine to which family a plant belongs. Once the family is determined turn to the write-up as designated in the key. In some cases, the family is large and additional keys will be necessary to determine the species.

This chapter is divided into the following sections:

    I. FLOWERS WITHOUT SEPALS AND PETALS; SEPALS USUALLY PETAL-LIKE, p. 112.
    II. FLOWERS WITH BOTH SEPALS AND PETALS; FLOWERS ASYMMETRICAL, FREE-PETAL, p. 113.
    III. FLOWERS WITH BOTH SEPALS AND PETALS; FLOWERS ASYMMETRICAL, UNITED-PETAL, p. 114.
    IV. FLOWERS WITH BOTH SEPALS AND PETALS; FLOWERS SYMMETRICAL, FREE PETAL, page 114.
    V. FLOWER WITH BOTH SEPALS AND PETALS; FLOWERS SYMMETRICAL, UNITED-PETAL, page 117.

I. FLOWERS WITHOUT BOTH SEPALS AND PETALS; SEPALS USUALLY PETAL-LIKE.

1. Leaves compound. Leaflets in groups of 3 or 4, leaflets circular. Male flowers appearing like tassels.

    See MEADOWRUE
    Buttercup family RANUNCULACEAE p. 168.

1. Leaves simple, basal, alternate or opposite.

    2. Flowers lavender, emerging before leaves. Leaf-like bract below flower. Foliage and flower hairy.

    See PASQUE FLOWER
    Buttercup family RANUNCULACEAE, p. 167.

bract

2. Flowers white, pink, or purple. Leaves appearing before flowers.   *dicots*

   3. Sepals generally white or pink. Fruit 3-sided, hard-shelled. Leaves basal or alternate.

   > Go to Buckwheat family
   > POLYGONACEAE, p. 162.

   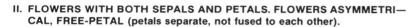

   3. Sepals showy, purple. Cluster of 2 to 6 flowers attached within a cup made of leaf-like bracts. Leaves opposite.

   > Go to Four-o'clock family
   > NYCTAGINACEAE, p. 155.

II. FLOWERS WITH BOTH SEPALS AND PETALS. FLOWERS ASYMMETRI‑CAL, FREE-PETAL (petals separate, not fused to each other).

1. Flowers resembling a sweet pea (papilionaceous); divided into 2 lateral petals (wings), 1 broad upper petal (banner), and 2 lower petals fused together (keel).

   > Go to Pea family
   > LEGUMINOSAE, p. 145.

   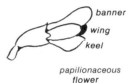

1. Flowers not papilionaceous.

   2. Flower parts in 5s.

      3. Lower flower petal with small sac. Leaves heart-shaped.

      > Go to Violet family
      > VIOLACEAE, p. 186.

      3. Lower petal without small sac. Sepals petal-like, upper ones hooded or spurred. Leaves palmately veined.

         4. Upper sepal spurred. Flowers white or blue.

         > See LARKSPUR
         > Buttercup family RANUNCULACEAE, p. 166.

         4. Upper sepal hooded. Flowers helmet-shaped, blue.

         > See MONKSHOOD
         > Buttercup family RANUNCULACEAE, p. 165.

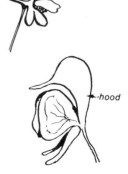

113

2. Flower parts in 4s. Petals bright yellow. Leaves finely dissected.

>See GOLDEN SMOKE
>Fumitory family FUMARIACEAE, p. 137.

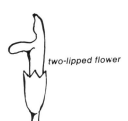

III. FLOWER WITH BOTH SEPALS AND PETALS; FLOWERS ASYMMETRICAL, UNITED-PETAL (flower parts fused).

1. Stems square. Leaves opposite.

    2. Flowers definitely two-lipped, in dense spherical clusters or loose clusters with bracts. Foliage aromatic.

    >Go to Mint family
    >LABIATAE, p. 142.

    2. Flowers very slightly two-lipped or asymmetrical, in spikes or loose heads.

    >Go to Vervain family
    >VERBENACEAE, p. 185

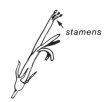

1. Stems round. Leaves alternate, opposite, or circled around the stem.

    3. Leaves alternate. Flowers scarlet. Stamens 5, united into a protruding tube.

    >See SCARLET LOBELIA
    >Bellflower family CAMPANULACEAE, p. 126.

    3. Leaves alternate or opposite. Flowers variously shaped and colored.

    >Go to Figwort family
    >SCROPHULARIACEAE, p. 174.

IV. FLOWERS WITH BOTH SEPALS AND PETALS: SYMMETRICAL, FREE-PETAL (petals not fused to each other, each petal separate).

1. Flowers with 4 petals.

    2. Stamens 8, stigma with 4 lobes. Sepals directed backward.

    >Go to Evening primrose family
    >ONAGRACEAE, p. 156.

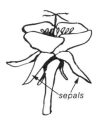

dicots

2. Stamens 6.

    3. Stamens 6, with 4 long and 2 short, not extending beyond the petals. Petals 4, in the shape of a cross. Fruit a round or elongated pod.

            Go to Mustard family
            CRUCIFERAE, p. 131.

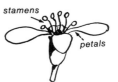

    3. Stamens 6, all the same length and extending beyond the petals (exerted). Plant strong-smelling.

            Go to Caper family
            CAPPARIDACEAE, p. 127.

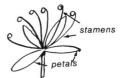

1. Flowers with 5 petals.

    4. Stamens many, more than 10.

        5. Leaves mostly basal, pinnate to lyre-like. Petals yellow, sepals russet-pink. Fruit a feathery plume.

            See OLD-MAN'S WHISKERS
            Rose family ROSACEAE, p. 170.

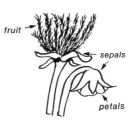

        5. Leaves not mostly basal. Flower various colored, sepals green.

            6. Leaves opposite with translucent or dark-colored dots. Flowers yellow.

                Go to St. Johnswort family
                GUTTIFERAE, p. 140.

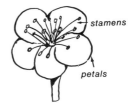

            6. Leaves alternate, simple or compound.

                7. Leaves alternate and compound. Bracts beneath the flowers. Stamens on the edge of a cup formed by the sepals and petals.

                    Go to Rose family
                    ROSACEAE, p. 169.

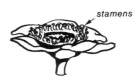

                7. Leaves alternate, simple.

                    8. Leaves pinnately lobed, covered with rough-barbed hairs, sticky. Flowers creamy-yellow, blooming late in the afternoon.

                      Go to Blazing star family
                      LOASACEAE, p. 152.

dicots

8. Leaves palmately veined, not covered by rough barbed hairs.

   9. Stamens on a cone. Flowers yellow. Leaves often dissected.

   > See BUTTERCUP
   > Buttercup family RANUNCULACEAE, p. 166.

   9. Stamens united into a tube. Flowers white, orange, or red.

   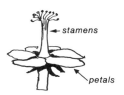

   > Go to Mallow family
   > MALVACEAE, p. 153.

4. Stamens 5 to 10, but not more than 10.

   10. Leaves compound.

       11. Leaves with 3 leaflets (shamrock-like). Flowers purple on a leafless stalk, closing when cloudy or at dusk.

       > Go to Oxalis family
       > OXALIDACEAE, p. 159.

       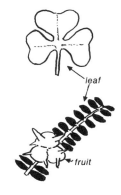

       11. Leaves pinnately compound. Fruit with stout spines. Flowers yellow.

       > Go to Caltrop family
       > ZYGOPHYLLACEAE, p. 187.

   10. Leaves simple.

       12. Leaves opposite. Stems swollen at the nodes. Anthers often colored. Flowers white.

       > Go to Pink family
       > CARYOPHYLLACEAE, p. 128.

       12. Leaves alternate or basal, if opposite only a few on the upper stems.

           13. Leaves palmately veined, dissected. Styles united, stigmas 5. Fruit beaked. Flowers white, pink, or purple. Upper leaves opposite, lower basal.

           > Go to Geranium family
           > GERANIACEAE, p. 139.

116

dicots

13. Leaves either linear or rounded, not dissected.

    14. Leaves linear. Plants stiffly erect. Flowers blue or yellow, falling off easily.

        Go to Flax family
        LINACEAE, p. 151.

    14. Leaves basal. Flowers small, greenish white, on leafless stalk.

        Go to Saxifrage family
        SAXIFRAGACEAE, p. 173.

V. FLOWERS WITH BOTH SEPALS AND PETALS; SYMMETRICAL, UNITED-PETAL (petals fused into a tubular, dish-shaped or funnel-formed structure termed a corolla).

1. Flowers in umbrella-like clusters.

    2. Plants with milky juice. Flowers in the shape of a crown, white to greenish-white.

        Go to Milkweed family
        ASCLEPIADACEAE, p. 122.

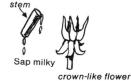

crown-like flower

    2. Plants without milky juice.

        3. Leaves opposite.

            4. Stems hollow. Leaves generally finely dissected or notched. Flowers small; white or yellow. Leafstalk (petiole) surrounding the stem.

                Go to Carrot family
                UMBELLIFERAE, p. 181.

umbrella-like cluster

            4. Stems not hollow. Leaves entire. Flowers like crowns, orange to red-orange.

                See BUTTERFLYWEED
            Milkweed family ASCLEPIADACEAE, p. 123.

        3. Leaves basal. Flowers tiny, white.

            See ROCK JASMINE
        Primrose family PRIMULACEAE, p. 164.

1. Flowers not in umbrella-like clusters.

5. Petals with four lobes. Stems square. Leaves in a circle around the stem. Flowers white.

> Go to Madder family
> RUBIACEAE, p. 172.

5. Petals and sepals with 5 or more lobes.

   6. Plant with milky juice. Leaves opposite. Stems forking. Flowers pink.

   > Go to Dogbane family
   > APOCYNACEAE, p. 121.

   opposite leaves

   6. Plant without milky juice. Leaves various.

      7. Plant twining, trailing, or vine-like.

         8. Large robust trailing plant. Flowers yellow, either male (with stamens) or female (with a pistil). Flowers of both sexes on 1 plant. Stamens 3, twisted. Fruit a gourd.

         > Go to Gourd family
         > CUCURBITACEAE, p. 135.

         8. Delicate trailing plant. Flowers funnel-like, plaited when in bud. Often with milky juice.

         > Go to Morning-glory family
         > CONVOLVULACEAE, p. 130.

         plait

         funnel-like flower

      7. Plants not twining, trailing or vine-like.

         9. Flower clusters twisted like a scorpion tail.

            10. Stamens extending beyond the petals. Leaves and stems covered with sticky hairs.

            > Go to Waterleaf family
            > HYDROPHYLLACEAE, p. 141.

            10. Stamens not extending beyond the petals. Leaves and stems covered with stiff hairs.

            > Go to Forget-me-not family
            > BORAGINACEAE, p. 124.

         9. Flower clusters not twisted like a scorpion tail.

            11. Flowers funnel-shaped, or like a long tube with a flattened top (salver-like).

dicots

12. Flowers white, funnel-like, large, over 8 cm wide. Fruit a spiny ball. Plant robust.

        See SACRED DATURA
Nightshade family SOLANACEAE, p. 179.

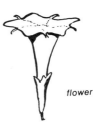
flower

12. Flowers funnel-like or salver-like, but small, yellow, red or blue.

   13. Flowers red or blue. Stigmas 3.

        Go to Phlox family
POLEMONIACEAE, p. 161.

salver-like flower

   13. Flowers yellow.

        See PUCCOON
Forget-me-not family BORAGINACEAE, p. 125.

11. Flowers not funnel-like or salver-like.

   14. Leaves at base of plant.

      15. Flowers tiny in dense spikes, petals 4, papery.

        Go to Plantain family
PLANTAGINACEAE, p. 160.

habit

      15. Flowers not in dense spikes, petals 5.

         16. Petals roundish, appearing separate, white to pinkish. Leaves leathery.

        See PIPSISSEWA
Heath family ERICACEAE, p. 136.

         16. Petals projecting backward (reflexed). Stigma and stamens forming a black point. Petals bright pink-purple.

        See SHOOTING STAR
Primrose family PRIMULACEAE, p. 164.

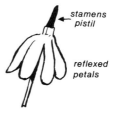
stamens pistil

reflexed petals

14. Leaves alternate or opposite, but not basal.

   17. Leaves alternate

      18. Flowers nodding.

        19. Flowers bell-shaped, blue. Basal leaves often different from the stem leaves.

        Go to Bellflower family
CAMPANULACEAE, p. 126.

bell-shaped flower

19. Flower tubular, yellow or blue.

> Go to Forget-me-not family
> BORAGINACEAE, p. 124.

tubular flower

18. Flowers not nodding. White or blue. Plait between the petals. Leaves and stems often with stiff hairs. Fruit a berry.

> Go to Nightshade family
> SOLANACEAE, p. 179.

plait

17. Leaves opposite or whorled.

20. Flowers either solitary or arranged along the stem. Wheel-shaped or funnel-like. Petals purple or yellow-green. Often plaited when in bud.

> Go to Gentian family
> GENTIANACEAE, p. 138.

20. Flowers in spikes, heads, or flat-topped clusters. Corolla often slightly asymmetrical.

21. Flowers in spikes or loose heads. Bracts with sepals. Corolla slightly 2-lipped. Stamens 4. Flowers purple or rose-purple.

> Go to Vervain family
> VERBENACEAE, p. 185.

flat-topped cluster

21. Flowers in flat-topped clusters. Pink to light pink. Plant smells like dirty feet.

> Go to Valerian family
> VALERIANACEAE, p. 184.

Dogbane Family APOCYNACEAE

Members of the dogbane family have milky juice and small cylindrical flowers with united petals. Flower parts are in fives. Five stamens are attached alternately to the lobes; the anthers adhere to a viscid material on the stigma. Leaves are opposite on the stem and droop gracefully.

I. FLOWERS PINK WITH REDDISH STRIPES.

Stems to 3 ft. Leaves oval to lance-shaped. Flowers bell-shaped or cylindrical, nodding; pink with reddish stripes. The plant is sometimes called WANDERING MILKWEED or MILK IPECAC. As are many other members of this family, SPREADING DOGBANE is poisonous. Generally found in disturbed areas, on dry gravelly slopes, and in sandy soils.

flower

SPREADING DOGBANE
Apocynum androsaemifolium
Greek: *apo*, away from; *kynos*, dog
Latin: *androsaemum*, like androsace
folium, leaf

II. FLOWERS WHITE.

Tall, erect plant, in clumps, growing to 3 ft. Leaves lance-shaped. Flowers to 5 cm long, white, cylindrical to urn-shaped. Stems used for making rope. Found along roadsides of the pinyon-juniper woodland.

flower

INDIAN HEMP
Apocynum cannabinum
Latin: *cannabis*, hemp

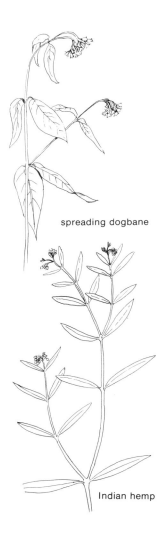
spreading dogbane

Indian hemp

121

Milkweed Family ASCLEPIADACEAE

Milkweeds are so named because the stems exude a milky juice or latex when injured. The flowers, which have parts in fives, look like miniature crowns. This appearance is created by an outgrowth of the corolla and filaments, forming the crown-like structure, or corona. In some species the petals project backward (are reflexed), and in others they stand upward. The stigma is five-angled; each of the five stamens adheres to the stigma through a waxy mass of pollen. This mass also unites the adjacent stamens. Flowers generally are in umbrella-like clusters. The fruits, called *follicles*, are filled with many cottony seeds. Although plants of some species are used medicinally, those of others contain poisonous alkaloids. Economically important members of the family include such ornamentals as butterflyweed and bloodflower, also houseplants such as star-fish flower and wax plant. The sap of a related species, *Matelea*, has been used as an arrow poison. Other relatives such as the Ceylon milk plant have been used as food.

I. FLOWERS GREENISH-WHITE, HOODS PURPLE-BROWN, PLANT RE— CLINING ON THE GROUND.

Stems reclining on the ground, to 2 ft long. Leaves alternate to almost opposite, linear to lance-shaped. Flowers with greenish white lobes; hoods brownish purple. The plant is called *inmortal* by the Spanish, who have used it medicinally. Found in the pinyon-juniper woodland.

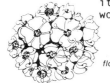
flowers

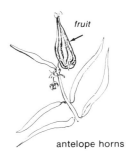
fruit

antelope horns

ANTELOPE HORNS
Asclepias asperula subsp. *asperula*
Greek: *Asklepios*, the god of medicine
Latin: *asper*, rough

II. FLOWERS PINK TO GREEN—PURPLE, STEMS ERECT.

Stems to 3 1/3 ft, hairy. Leaves opposite on the stem, oblong to oval lance-shaped. Flower bearing a horn; lobes of sepals and petals directed backward. Found in disturbed soils.

flower

SHOWY MILKWEED
Asclepias speciosa
Latin: *speciosus*, showy

Stems to 3 ft. Leaves rounded-oval. Flowers greenish white to greenish yellow; hoods yellow-white. Disturbed soils.

BROAD-LEAVED MILKWEED
Asclepias latifolia
Latin: *latus*, side; *folium*, leaf

showy milkweed

III. FLOWERS WHITE. STEMS ERECT; LEAVES, NARROW, IN A CIRCLE AROUND THE STEM.

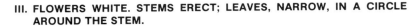
milkweed

Stems to 2 1/2 ft. Leaves narrowly linear, in a circle around the stem. Flowers white to cream-colored, sometimes greenish purple, lobes turned down. This plant is poisonous and has been incriminated in loss of livestock. It is an invader of disturbed areas.

POISON MILKWEED
Asclepias subverticillata
Latin: *sub-*, below; *vertex*, a turning

poison milkweed

IV. FLOWERS ORANGE, STEMS ERECT.

Stems to 1 1/2 ft, erect, lacking milky sap. Leaves linear to lance-shaped; margins often rolled under. The flowers in umbrella-like clusters at the top of the plant. Butterflies are attracted to the plant, giving it the common name BUTTERFLYWEED. Canyons of the pinyon-juniper woodland and mixed conifer forest.

BUTTERFLYWEED
Asclepias tuberosa
Latin: *tuber*, a swelling

flower

butterflyweed

Forget-me-not Family BORAGINACEAE

Flowers of most members of this family are small and colored white, blue, or yellow. Flower clusters are often shaped like a coiled scorpion-tail. Leaves and stems are covered with stiff hairs. The fused petals of the flowers may be tubular, salverform, or funnel-like. Leaves are simple and alternate on the stem. Seeds have four nutlets which in some species such as the STICKSEED are armed with barbed prickles. Economically the family is of little importance except for such ornamentals as heliotrope, Virginia bluebells, forget-me-not, lungwort, hound's tongue, comfrey, and viper's bugloss.

I. FLOWERS TUBULAR, NODDING, BLUE.

BLUEBELLS
Mertensia spp.
Honors Francis C. Mertens
German botanist

Species commonly found in the area include:

CHIMINGBELLS *M. lanceolata*. Stems to 1 ft. Leaves lance-shaped, hairy on the upper surface, smooth below. Lower leaves with a leafstalk, upper leaves sitting against the stem. Flowers tubular, nodding, to 9 mm long. Found in ponderosa pine and mixed conifer forests.

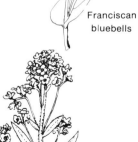

chimingbells

flower

FRANCISCAN BLUEBELLS *M. franciscana*. Stems to 3 ft. Leaves at the base of the plant large, to 10 cm long, lance-shaped to oblong-oval. Flowers white to blue, sometimes pinkish, to 9 mm long. Found along streams.

Franciscan bluebells

II. FLOWERS SALVER-LIKE; BLUE OR WHITE.

1. Plant growing to 3 ft.

Stems growing to 3 ft tall, covered with hairs pressed downward; base of the hairs white or enlarged. Leaves lance-shaped. Flowers to 2 cm wide, white to blue. Seeds with prickles. Found in ponderosa pine and mixed conifer forests.

STICKSEED FORGET-ME-NOT
Hackelia floribunda
Honors P. Hackel, German professor of agriculture
Latin: *flos*, flower; *abundare*, to overflow

seed

forget-me-not

1. Plant growing to between 1 and 2 ft.

Stems to 2 ft. Flowers to 3 mm, blue to pale blue. Seeds with bristles or barbs. Plant most noticeable for the barbed seeds, which catch clothing or animal fur. Found along roadsides, trails, and in disturbed soils.

STICKSEED
Lappula redowskii
Honors D. Redowsky, 19th-century
Russian botanist
Latin: *lappula*, small bur

forget-me-not

stickseed

III. FLOWERS WHITE, TINY.

Stems to 1 ft, covered with stiff, short hairs. Leaves linear to lance-shaped, likewise covered with hairs. Flowers white, 3 mm long. Found in disturbed soils.

JAMES HIDDENFLOWER
Cryptantha jamesii
Greek: *kryptos*, hidden; *anthos*, flower
Honors Dr. Edwin James (1797-1861)
American physician and botanist

James hiddenflower

IV. FLOWERS FUNNEL-LIKE, YELLOW.

PUCCOON
Lithospermum spp.
Greek: *lithos*, stone; *sperma*, seed

Plants of this genus were used as medicine and food by various tribes throughout the West. The fleshy taproot of some species contains a purple pigment which was used as a dye. Species commonly found in the area include:

PUCCOON *L. multiflorum*. Stems growing to 2 ft tall, covered with stiff hairs. Leaves linear to linear-lance-shaped, becoming bract-like at the top. Flower funnel-like, yellow or orange, to 1 cm long. Found in the ponderosa pine and mixed conifer forests.

FRINGED PUCCOON *L. incisum*. Stems growing to 2 ft. Leaves linear to lance-shaped. Flowers yellow, to 2.5 cm long, petal margins fringed. Found in pinyon-juniper woodland and ponderosa pine forest.

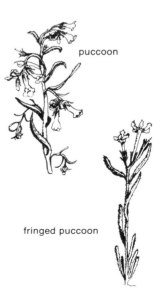

puccoon

fringed puccoon

125

Bellflower Family CAMPANULACEAE

Plants of the bellflower family have symmetrical, bell-shaped flowers (HAREBELL) or asymmetrical, two-lipped flowers (LOBELIA). All flowers have five sepals and five petals. Stamens are attached alternate to the petals and are often united. Leaves are alternate. Plants often have milky or acrid juice. Domestic bellflowers, lobelia, and a number of other ornamentals occur in this family.

I. FLOWERS RED, ASYMMETRICAL.

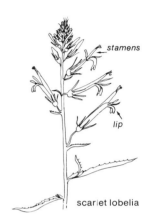

A striking plant growing to 3 ft. Flower characterized by a 2-lipped corolla, the upper lip with 2 lobes, the lower with 3 large lobes. Stamens protrude through the upper 2 lobes. Various species of lobelia have alkaloids which make them poisonous. Found around the springs in White Rock Canyon. Particularly beautiful in the autumn.

SCARLET LOBELIA
Lobelia cardinalis
Honors Matthias de l'Obel, 16th-century Flemish herbalist
Latin: *cardo*, hinge (i.e. something on which an important thing depends)

II. FLOWERS BLUE, SYMMETRICAL.

BLUEBELLS
Campanula spp.
Latin: *campanula*, little bell

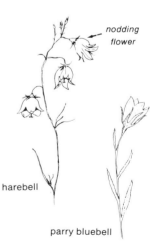

HAREBELL *C. rotundifolia* Grows to 1 1/2 ft. Stems unbranched, delicate, erect. Leaves basal, oval to heart-shaped but wither early, seldom evident. Upper leaves linear. Flowers blue-violet, bell-shaped, nodding. Found in moist canyons, mixed conifer forests, and high meadows.

PARRY BLUEBELL *C. parryi* Usually grows to 1 ft. Stems 1-flowered. Lower leaves spatula-like; upper leaves linear. Flowers purple, not nodding, to 2 cm long. Found in woods and canyons of the mixed conifer forests.

Caper Family CAPPARIDACEAE

Members of the caper family have four petals, four sepals and six stamens which extend beyond the petals (exerted). The white or purple petals are narrowed at the base (clawed) and the flowers are in clusters at the end of the branches. These plants have strong-smelling, watery juice, giving one member the charming name of STINKWEED. Capers, used in seasoning foods, are the dried flower buds of a Mediterranean species. Spider flower is a popular garden annual. Two species occur in the area. Both are found in disturbed or sandy soils of roadsides, trails, arroyos, and river banks of the lower elevational zones.

I. FLOWERS PURPLE

Stems to 5 ft. Leaves with 3 lance-shaped to oblong leaflets. Flowers purple with petals narrowed at the base and stamens extending beyond petals. This species has been of importance to the cultures of the Southwest. Young shoots have been used as a potherb, freshly-cut leaves to reduce inflammation from insect bites, and the hardened residue from the boiled plant to paint pottery.

STINKWEED, ROCKY MOUNTAIN BEE-PLANT
Cleome serrulata
Greek: *kleome*, a mustard-like plant
Latin: *serra*, a saw

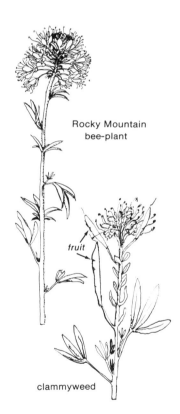

Rocky Mountain bee-plant

II. FLOWERS WHITE TO YELLOWISH.

Stems to 2 ft. Stems and leaves sticky and strongly scented. Petals white to yellowish, narrowed at the base. Stamens with purple filaments.

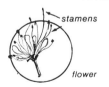

CLAMMYWEED
Polanisia trachysperma
Greek: *polys*, many; *anisos*, unequal
tachys, rough; *sperma*, seed

clammyweed

Pink Family CARYOPHYLACEAE

Plants in the pink family are usually small and rather delicate, with white, pink, or purple flowers. There are five sepals, five petals, and ten stamens, often with colored anthers. Petals are often narrowed at the base (clawed) and sometimes notched at the end. Stems are enlarged at the nodes. Leaves, linear or lance-shaped in outline, are opposite on the stem. Economically the family is important for several ornamentals, including the florist's carnation and baby's-breath. There are a large number of species in the area.

I. FLOWERS PURPLE, NODDING. SEPALS WITH 10 RIBS.

Stems to 1 1/2 ft, covered with sticky hairs. Lower leaves linear to lance-shaped. Petals purple, deeply 2-lobed. Found in high meadows and canyons.

CATCHFLY
Silene scouleri
Greek: *Silenus*, the mythical leader
of the satyrs and tutor of Bacchus
Honors Dr. John Schouler (1804-1871)
Scottish naturalist

catchfly

II. FLOWERS WHITE.

1. Petals notched.

Stems to 1 ft, covered with hairs directed backward. Leaves linear to narrowly oblong. Flower petals white, notched, to 7 mm; sepals covered with sticky hairs. Found in meadows and ponderosa pine and mixed conifer forests.

mouse-ear chickweed

MOUSE-EAR CHICKWEED
Cerastium arvense
Greek: *keras*, horn
Latin: *arvum*, field

pink

Stems to 1 1/2 ft, sticky-hairy, sharply angled. Leaves linear to lance-shaped. Petals white, 2-lobed, to 8 mm; longer than sepals. Sepals with papery margins. Found in high-elevation forests and meadows.

flower

JAMES STARWORT
Stellaria jamesiana
Latin: *stella*, star
Honors Edwin James (1797-1861)
American physician and botanist

James starwort

Fendler's starwort

1. Petals not notched.

Stems to 2 ft, sticky-hairy. Leaves filament-like, sharp-pointed. Sepals papery margined. Petals to 5 mm, shorter than the sepals or just as long. Found in meadows and forests of the ponderosa pine and mixed conifer forests.

FENDLER'S SANDWORT
Arenaria fendleri
Latin: *arena*, sand
Honors August Fendler (1813-1883)
German-born naturalist and explorer

Morning-glory Family CONVOLVULACEAE

Plants of the morning-glory family may be annual or perennial herbs, often twining. DODDER (see p. 48), a non-showy member, is a leafless twining parasite on herbaceous plants. Leaves are alternate on the stem, either arrow-shaped or heart-shaped. In the showy species flowers are funnel-like or tubular, and twisted in the bud. Except for the pistil, the flower parts are in fives. Economically important members include the sweet potato, morning glory, cypress vine, and wood rose.

Species found in the area are vines or twining plants (see also p. 12). They include:

I. FLOWERS WHITE.

Prostrate trailing plant. Flowers white, funnel-shaped. Extremely noxious weed of gardens and lawns. Hard to eradicate once established. (See also p. 15.)

FIELD BINDWEED
Convolvulus arvensis
Latin: *convolvere*, to entwine
arvus, field

field bindweed

II. FLOWERS RED.

Herbaceous plant twining on vegetation. Flowers bright red.

STAR-GLORY
Ipomoea coccinea
Greek: *ipos*, worm; *kokkinos*, scarlet

star-glory

Mustard Family CRUCIFERAE

Members of the mustard family can be distinguished by their flowers, which have four sepals, four petals, and six stamens, four long and two short. The petals appear to form a cross (crucifix), from whence comes the family name CRUCIFERAE. The leaves, variously shaped, are alternate. The fruits are specialized and are termed *silicles* or *siliques*. The family contains many weedy species, generally having small white or yellow flowers; a few representatives, however, are showy. Because some of the genera are large, with many representatives in the area, only generic descriptions are given here. Many important garden vegetables, such as cabbage, cauliflower, turnip, and radish, are members of the family.

This section is divided as follows:
   I. PLANT OF STREAMS AND SPRINGS. FLOWERS WHITE, p. 131
  II. PLANT OF DRY SOILS. FLOWERS YELLOW OR WHITE, p. 132
 III. PLANT OF DRY SOILS. FLOWERS PINK TO PURPLE, p. 133
 IV. PLANT OF DISTURBED SOILS, ROADSIDES AND GARDENS. FLOWERS SMALL, p. 133

I. PLANTS OF STREAMS AND SPRINGS. FLOWERS WHITE.

flower

Stems submerged in water. Leaves pinnately divided, oval or with segments, the terminal one larger than the lateral ones. Flowers white. The leaves have a mildly peppery taste and are used all over the world as a salad green and soup ingredient. One should not eat the plant raw if taken from the wild. Many local waters have been contaminated with a protozoan parasite, *Giardia* spp., which causes a gastric disturbance, giardiasis. The symptoms range from mild to severe. Plant naturalized from Europe. Found in streams and springs.

watercress

                  WATERCRESS
         Rorippa nasturtium-aquaticum
     Latin: *ros*, dew; *ripa*, bank
   nasus, nose; *torquere*, to twist
                  aqua, water

fruit

Stems growing to 2 ft. Leaves heart-shaped. Margins toothed or wavy. Flowers white, to 1 cm long. Like watercress it can be used in salads, although the caution given above concerning possible contamination also applies. Found along streams and in springs.

flowers

bittercress

         HEARTLEAF BITTERCRESS
           Cardamine cordifolia
Greek: *kardamon*, a kind of cress
chorde, string. Latin: *folium*, leaf

II. PLANTS OF DRY SOILS. FLOWERS YELLOW OR WHITE.

1. Flowers yellow.

Stems to 2 ft. Leaves narrowly linear to lance-shaped; margins smooth to toothed. Flowers yellow, 1.5 to 2 cm across. Fruits erect. The plant has been used medicinally to treat pain, sunburn, and pneumonia.

WESTERN WALLFLOWER
Erysimum capitatum
Greek: *eryo*, to draw blisters
(from *erythros*, red). Latin: *caput*, head

Stems growing to 20 cm. Basal leaves spatula-like to lance-shaped. Stem leaves oval to lance-shaped, covered with branched hairs. Flowers pale to bright yellow, in clusters at the top of the stem.

WHITLOW GRASS
Draba aurea
Greek: *drabe*, name applied by Dioscorides to some form of cress
Latin: *aurum*, gold

flower

1. Flowers white.

Stems to 1 ft, often growing in clumps. Leaves oval to spatula-like, arranged in a rosette at the base of the plant. Stem leaves clasping the stem, with ear-like lobes at the base. Flowers white tinged with purple, to 4 mm long, in long clusters at the top of the stem. Fruits flat, rounded and notched at the top.

PENNYCRESS, WILD CANDYTUFT
Thlaspi alpestre
Greek: *thlao*, to crush (from flattened pods)
Latin: *alpestris*, of mountains

Stems growing to 2 ft. Leaves linear to lance-shaped; margins wavy. Petals white, to 8 mm, clustered at the end of the stem. Fruits flattened and resemble spectacles. Sandy areas along the Rio Grande.

SPECTACLE POD
Dithyrea wislizenii
Greek: *di*, two; *thryeos*, shield
Honors Frederick Wislizenus (1810-1899)
German-born physician and naturalist

fruit

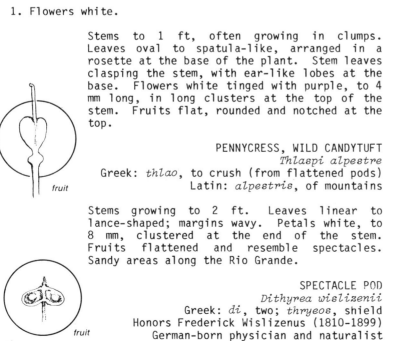
fruit

fruit
western wallflower
whitlow grass
wild candytuft
spectacle pod

III. PLANTS OF DRY SOILS. FLOWERS PINK TO PURPLE.

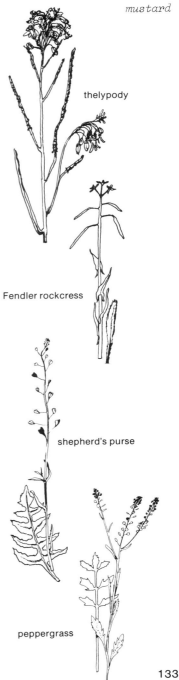

mustard

Stems growing to 4 ft. Leaves at the base pinnatifid, lyre-shaped. Stem leaves linear to lance-shaped. Flowers purple to whitish, the petals narrowed at the base. Fruits rounded siliques.

flowers

THELYPODY
Thelypodium wrightii
Greek: *thelys*, female; *podos*, foot
Honors Charles Wright (1811-1885)
American botanist

thelypody

Stems to 2 ft, covered with branched hairs. Leaves lance-shaped, in a rosette at the base. Upper leaves clasping the stem with ear-shaped lobes. Flowers pink to purple, 5 to 8 mm long. Fruit a silique.

FENDLER ROCKCRESS
Arabis fendleri
Greek: *Arabis*, Arabia
Honors August Fendler (1813-1883)
German-born botanist

Fendler rockcress

IV. PLANTS OF DISTURBED SOILS, ROADSIDES, AND GARDENS. FLOWERS SMALL.

1. Flowers white.

flower

fruit

Stems to 1 1/2 ft. Leaves basal, pinnate to lyre-shaped. Flowers white, to 2 mm wide, in long clusters at the top of the stem. These are very early spring bloomers. One is more apt to see the small triangular fruits, 8 mm long. These resemble the old shepherd's purse; thus the common name.

SHEPHERD'S PURSE
Capsella bursa-pastoris
Latin: *capsella*, little box
bursa, purse; *pastor*, shepherd

shepherd's purse

Stems to 1 1/2 ft. Leaves pinnate. Flowers small, white. The plant has a peppery taste. The seeds can be used for seasoning soups and salads.

flower

fruit

PEPPERGRASS
Lepidium medium
Greek: *lepis*, a scale

peppergrass

133

1. Flowers yellow.

Stems to 2 1/2 ft. Leaves mostly at the base of the plant, linear to lance-shaped, covered with star-shaped hairs. Margins smooth or toothed. Flowers yellow. Fruits inflated, shaped like a ball, popping when stepped on.

BLADDERPOD
Lesquerella intermedia
Honors Leo Lesquereux (1805-1889)
Swiss-born botanist
Latin: *intermedius*, intermediate

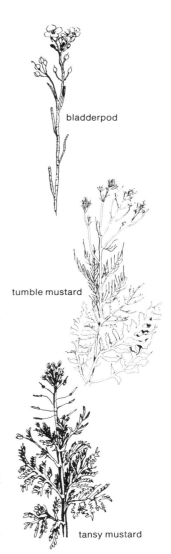

bladderpod

tumble mustard

tansy mustard

Stems to 3 ft, much-branched, breaking off at maturity to become a tumbleweed. Leaves pinnatifid. Flowers yellow to dark cream-colored. Seed pods long and narrow. The plant produces large quantities of seeds which are dislodged and scattered when the plant tumbles. Seeds were introduced in hay and grain along the Great Northern Railroad, and the species has since become widespread.

TUMBLE MUSTARD
Sisymbrium altissimum
sisymbrium was ancient Greek name for the mustard family
Latin: *altissimum*, very high

Stems to 3 ft. Leaves pinnate. Flowers bright yellow, to 2 mm. Fruits narrowly linear. Indians used the seeds of the tansy mustard for making pinole. Cattle ranging in southern New Mexico may get "paralyze tongue" by eating the plant.

flower

TANSY MUSTARD
Descurainia richardsonii
Honors Francois Descurain (1658-1740)
French apothecary, and
Sir John Richardson (1787-1865)
English naturalist

Gourd Family CUCURBITACEAE

Plants of the gourd family are large, sprawling vines. The bright yellow flowers are unisexual, having either pistils or stamens, but not both. Male flowers have three stamens with twisted anthers; the ovaries of the female flowers produce a swelling beneath the petals. Leaves are large, alternate on the stem, and somewhat triangular in outline. Toward the fall the plants produce small gourds. Economically important members of the family include pumpkin, muskmelon, and cucumber. Wild species inhabit disturbed soils along roadsides, riverbanks, or railroads.

An odoriferous, sprawling plant growing to 3 ft across. Leaves large, triangular. Flowers yellow. See VINES AND TRAILING PLANTS, p. 16.

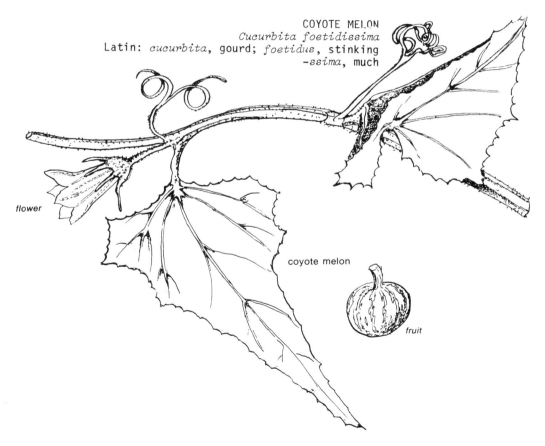

COYOTE MELON
Cucurbita foetidissima
Latin: *cucurbita*, gourd; *foetidus*, stinking
-*ssima*, much

Heath Family ERICACEAE

The heath family is a large, diverse group including such shrubs as BEARBERRY and WHORTLEBERRY (p. 29) and a number of herbaceous representatives. Rhododendrons, blueberries, trailing arbutus, Scotch heather, huckleberries, and cranberries are family members of economic importance. Some heaths--PINEDROPS and PINESAP--(p. 50) are saprophytes and lack chlorophyll. The flowers in some species have sepals, petals, and stamens partially fused; in others the flowers appear to be free-petalled. The most common herbaceous member of the heath family found in the area is PYROLA.

I. FLOWERS IN A ONE-SIDED CLUSTER.

Plant with woody stems, growing to 20 cm tall. Leaves basal; margins finely toothed. Sepals oval. Petals elliptical to oblong, greenish white. Found in shaded areas of the ponderosa pine and mixed conifer forests.

SIDEBELLS
Ramischia secunda
Ramischia, origin unknown
Latin: *secundus*, following

sidebells

II. FLOWERS NOT IN A ONE-SIDED CLUSTER, URN-SHAPED.

Stems to 20 cm. Leaves elliptical to kidney-shaped or spherical. Margins with rounded teeth. Flowers greenish white, to 7 mm long; style bent. Shaded areas of the ponderosa pine and mixed conifer forests.

WINTERGREEN, PYROLA
Pyrola chlorantha
Latin: *pyrus*, pear (because the leaves are similar)
Greek: *chloros*, green; *anthos*, flower

pyrola

III. FLOWERS NOT IN ONE-SIDED CLUSTER, NOT URN-SHAPED.

Perennial growing to 20 cm tall. Leaves thick, shiny green, oblong lance-shaped to spatula-like, margins toothed. Flowers pinkish to white, to 6 mm long, nodding.

PIPSISSEWA
Chimaphila umbellata
Greek: *cheima*, winter; *philos*, loving
Latin: *umbella*, parasol

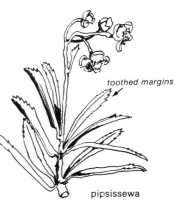
toothed margins

pipsissewa

Fumitory Family FUMARIACEAE

Members of the fumitory family have asymmetrical flowers and compound leaves. The garden ornamentals bleeding heart and Dutchman's breeches belong to this family. GOLDEN SMOKE is the only naturally-occurring representative found in the area.

An annual with stems to 1 1/2 ft, often prostrate. Leaves alternate, bluish green, dissected into narrow segments. Flowers with 2 sepals, falling easily. Petals 4; inner 2 petals fused, outer 2 spurred. Stamens united into 2 groups of 3 with winged filaments. In disturbed soils throughout the area.

GOLDEN SMOKE
Corydalis aurea
Greek: *korydalis*, crested lark
(from *korys*, helmet)
Latin: *aurum*, gold

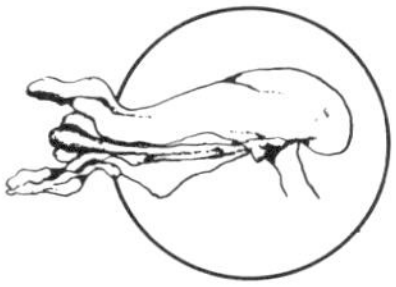

flower

golden smoke

Gentian family GENTIANACEAE

Gentians are usually plants of higher altitudes. Most members are low-growing herbs but one member, MONUMENT PLANT, can grow to seven feet. Leaves are opposite and attached directly to the stem. Flowers may be tubular, wheel-shaped, or salver-like; usually they have four or five lobes, but may have as many as twelve. Some members have fringed or pitted glands at the base.

I. PLANTS GROWING UP TO 7 FT TALL; FLOWERS WHEEL-SHAPED, GREENISH, WITH PURPLE DOTS.

Stems stout. Leaves large, linear to lance-shaped; some basal, others encircling the stem in groups of 3 to 7. Flowers positioned along the stem in axils of leaves; wheel-shaped, deeply parted, often dotted with purple. Two fringed glands on each lobe. It has been reported that the Indians ate the fleshy root of this species. The roots mixed with lard reportedly made a good insecticide. Found in canyons and on slopes in the mixed conifer forest.

DEER'S EARS, MONUMENT PLANT
Swertia radiata
Honors Emanuel Sweert, 16th-century Dutch botanist
Latin: *radius*, a spoke

II. PLANT TO 1 FT, FLOWERS BLUE-PURPLE, PLAITED.

flower

Stems clustered, growing to 1 ft tall. Leaves lance-shaped to oval. Flowers tubular with plaits between the petal lobes. Found in the ponderosa pine and mixed conifer forests.

PRAIRIE GENTIAN
Gentiana affinis
From Gentius, king of Illyria who first discovered the plant's medicinal properties
Latin: *affinis*, adjacent

III. PLANTS TO 15 CM, FLOWERS WHITE TO BLUE, FRINGE OF HAIRS AT THE THROAT OF THE FLOWER.

flower

Delicate annual growing to 15 cm. Leaves spatula-like to oblong lance-shaped. Flowers in the axils of the leaves and at the tip of the stem; each with 5 star-shaped lobes and a crown inside the lobes. Found in mixed conifer forests.

ROSE GENTIAN
Gentiana strictiflora
Latin: *strictus*, tight; *flos*, flower

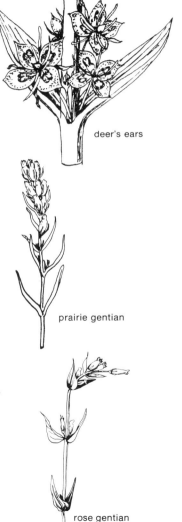

deer's ears

prairie gentian

rose gentian

Geranium Family GERANIACEAE

Geraniums have flower parts in five or multiples of five. Styles are united, fused at the base, but the five stigmas are separate. Leaves may be palmately lobed or dissected, with sticky hairs. Plants are often odoriferous. The elongated fruit has the appearance of a crane's or heron's bill, giving the common names CRANESBILL or HERONSBILL. The family contains the florist's geranium *Pelargonium zonale*. Hybrids have been developed for the production of aromatic oils.

I. LEAVES PINNATELY DIVIDED, FLOWERS PURPLE.

Stems trailing or lying on the ground, to 1 ft long. Leaves in rosettes at base of plant. Flowers in umbrella-like clusters, pink with dark veins of rose to purple. Sepals greenish, hairy, pointed, with bristle-like hairs at the tip. Fruit coiling when mature. This plant appears to have been introduced from Europe with the Conquistadores. Alfileria is from the Spanish word meaning "a pin." Teas were made from the root and used to treat bleeding and inflammations. One of the earliest and latest blooming plants. Found in disturbed soils, lawns, and along roadsides.

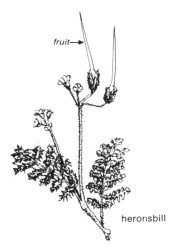
fruit→
heronsbill

ALFILERIA, HERONSBILL
Erodium cicutarium
Greek: *erodios*, heron
Latin: *cicuta*, hemlock

flower

II. LEAVES PALMATELY LOBED, FLOWERS PURPLE OR WHITE.

GERANIUM
Geranium spp.
Greek: *geranos*, crane

JAMES GERANIUM *Geranium caespitosum*. Grows to 1 1/2 ft. Flowers rose-purple, petals spreading backward; sepals bristle-tipped. Called by the Spanish *patita de leon* because the leaf looks like a lion's paw. Used as a gargle for sore throats and tonsilitis. Found in canyons as well as along roadsides of the ponderosa pine and mixed conifer forests.

RICHARDSON'S GERANIUM *Geranium richardsonii*. Grows to 1 1/2 ft. Flowers white to very pale purple, with purple veins; flower stem with purple, glandular hairs. Leaves with 5 to 7 lobes. Found in moist woods and canyons of the ponderosa pine and mixed conifer forests.

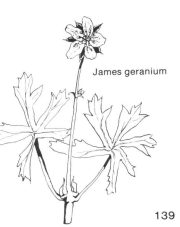
James geranium

Richardson's geranium

St. Johnswort Family GUTTIFERAE

The leaves of this family are distinctive. They are either opposite or in a circle around the stem, and have black or translucent dots. Flowers are large and bright yellow, with five sepals and petals. The numerous stamens are united by their filaments into bunches. Klamathweed *Hypericum perforatum* is a noxious range weed naturalized from Europe. Light-skinned cattle, horses, or sheep get a photosensitized dermatitis after eating the flowering stage of the plant. In order to eradicate the species, a beetle *Chripolina gemellata*, which feeds on the plant, was brought from Europe. The common name refers to a legend that the plant first blooms the 24th of June, St. John the Baptist's day.

Stems to 2 ft. Leaves oval to elliptical, with black dots. Flowers yellow, to 1.5 cm long with black dots on the margins of the petals. Sepals to 5 mm long with black dots on the margins. Found in canyons and on slopes of the mixed conifer forest.

ST. JOHNSWORT
Hypericum formosum
Greek: *hypo*, under; *ereike*, heath
Latin: *formosa*, shapely

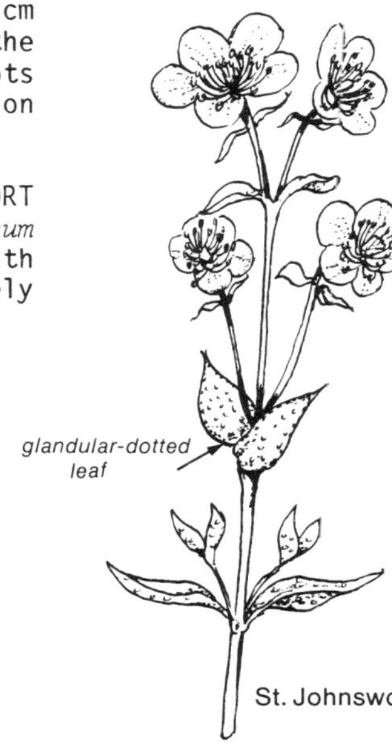

glandular-dotted leaf

St. Johnswort

Waterleaf Family HYDROPHYLLACEAE

Members of this family have glandular hairs that make the plants sticky or clammy. Flowers are in scorpion-tailed or coiled clusters. The fused petals are in fives. The five stamens which alternate with the corolla lobes extend beyond the petals, giving the plant a feathery appearance. Many members of this family are weedy; a few are showy.

I. PLANT GROWING TO 3 FT; ALONG STREAMS OR IN WET PLACES.

Stems growing to 3 ft, coarsely hairy. Leaves to 1 ft long, oval to oblong in outline, pinnatifid with 9 to 13 main segments, coarsely toothed. Flowers generally white but sometimes purplish; stamens extend beyond the petals. Young shoots and roots were used as a potherb by Indians. The leaves contain a high percent of water and the plant is named "washy feed" by stockmen. Found along streams.

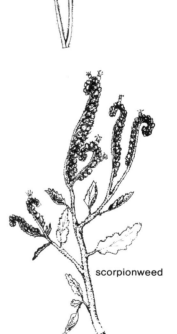
squaw lettuce

SQUAW LETTUCE
Hydrophyllum fendleri
Greek: *hydro*, water; *phyllon*, leaf
Honors August Fendler (1813-1883)
German-born naturalist and explorer

II. PLANT GROWING TO 2 FT; USUALLY DISTURBED OR DRY SOILS.

SCORPIONWEED
Phacelia spp.
Greek: *phakelos*, bundle

Stems to 2 ft. Leaves simple or pinnately compound. Margins often deeply lobed. Flowers white, blue or violet, funnel-like in compact clusters that coil and somewhat resemble a scorpion's tail. Some species are sticky and have a disagreeable onion-like odor. Species found in the area include:

flower

SCORPIONWEED *P. corrugata*. Flowers deep blue or violet.

SCORPIONWEED *P. heterophylla*. Flowers white to pink.

scorpionweed

Mint Family LABIATAE

Members of the mint family have asymmetrical flowers with fused petals, square stems, aromatic foliage, and fruit consisting of four nutlets. The sepals and petals of the strongly two-lipped flowers have five lobes. There may be two to four fertile stamens (those with anthers.) Flowers may be crowded in clusters around the stem or in spikes or clusters at the ends of stems. Leaves are simple and opposite. Economically the family is important as a provider of culinary herbs, aromatic oils, and garden ornamentals. Lavender is an important source of perfume; sage, basil, thyme and savory are important herbs; and salvia, bugloss, and lion's ears are important ornamentals.

I. FLOWERS DEEP PURPLE. PLANTS OF STREAMBANKS AND MOIST PLACES.

Stems growing to 1 ft; leafy, spreading or erect. Leaves somewhat heart-shaped with long leafstalks. Flowers violet-purple, in dense bracted spikes. Sepals and petals 2-lipped, the upper lip arched, the lower lip 3-lobed. This plant was formerly used to treat hemorrhages, for diarrhea, and as a gargle. Tea can be made from the leaves. Found along streams throughout the area.

selfheal

flower

SELFHEAL, HEALALL
Prunella vulgaris
German: *die Braune*, a disease plants of this genus were supposed to cure
Latin: *vulgaris*, common

II. FLOWERS LIGHT PURPLE, IN CLUSTERS AT TOP OF STEM. USUALLY IN MEADOWS OR ALONG STREAMS.

Erect perennial; stems to 2 ft tall. Leaves oval lance-shaped; upper surface hairy, lower covered with wool-like hairs. Leaf margins finely toothed. Flowers lavender-purple, 3 cm long, with stamens extending beyond the petals. The leaves have been used to make a tea and for flavoring cooked food. An antiseptic, thymol, is present in the volatile oil. Spanish New Mexicans used the tea to treat cough. The plant is also a source of vegetable dyes. Other common names are OSWEGO TEA, OREGANO, and BEEBALM. Found throughout the area.

horsemint

flower

HORSEMINT
Monarda menthaefolia
Honors Nicholas Monardes, 16th-century Spanish physician and botanist
Latin: *mentha*, mint; *folium*, leaf

III. FLOWERS PINK TO WHITE TINGED WITH PINK; PLANTS OF DISTURBED SITES.

mint

Stems to 1 ft, covered with downward-directed hairs. Leaves oblong to lance-shaped, surfaces smooth. Leaf margins finely toothed. Flowers to 1 cm, light rose-white, the lower lip often darker, spotted. This plant was used by Spanish New Mexicans for flavoring food and can be used as a substitute for oregano. Found in disturbed soils.

PONYMINT
Monarda pectinata
Latin: *pectin*, comb

ponymint

Stems to 2 ft. Leaves with spine-like teeth on the margins. Flowers light rose to bluish, to 1.5 cm long, in dense terminal clusters. Flower 2-lipped, the upper lip 3-lobed. A spine-like bract present beneath each flower. The Havasupai Indians of Arizona reputedly make a nutritious flour from the seeds; such flour was also prepared by early pioneers. Found in recently burned areas and other disturbed soils.

flower

DRAGONHEAD
Moldavica parviflora
Moldavia, a district in Rumania
Latin: *parvus*, small; *flos*, flower

dragonhead

Stems to 1 1/2 ft. Leaves oblong to lance-shaped; margins toothed. Flowers to 6 mm long, pink to light rose. Leaves and stems can be used for herbal teas. Found in moist places within canyons.

MINT
Mentha arvensis
Latin: *mentha*, mint; *arvus*, field

mint

IV. FLOWERS WHITISH. WEEDY PLANTS OF GARDENS, ROADSIDES, AND TRAILS.

Stems to 3 ft tall. Stems and leaves covered with white, woolly hairs. Flowers crowded into dense, somewhat spherical clusters in the leaf axils. The sepals have 10 spreading, hooked teeth and form a tubular calyx. Introduced from Europe. This plant is used to flavor candy and cough medicine; in Europe, horehound ale, flavored with horehound rather than hops, is popular.

flower

HOREHOUND
Marrubium vulgare
Latin: *marrubium*, horehound; *vulgaris*, common

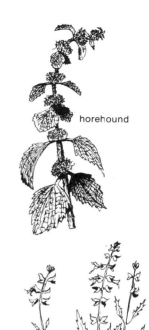
horehound

V. FLOWERS LIGHT PURPLE OR WHITE TINGED WITH BLUE. PLANTS OF FIELDS AND ROADSIDES.

Stems to 1 ft. Leaves oblong lance-shaped to linear. Leaf margins smooth, wavy, or slightly toothed. Flower to 1.2 cm long, 2-lipped. In California this plant is known as *chia*. The seeds were roasted and eaten; Indians made a pinole from them.

ROCKY MOUNTAIN SAGE, SALVIA
Salvia reflexa
Latin: *salvia*, sage
reflexa, bent backward

Stems growing to 25 cm. Leaves linear, elliptical, or oval, with smooth margins. Flowers to 7 mm long in axils of the leaves, purple to rose pink. Found on dry mesas and in rocky canyons.

FALSE PENNYROYAL
Hedeoma drummondii
Greek: *hedys*, sweet; *osme*, odor
Honors Thomas Drummond (1780-1835)
Scottish herbalist and explorer

flower

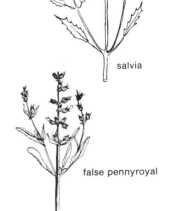
salvia

false pennyroyal

Pea Family LEGUMINOSAE

The flower of the pea family is distinctive, with its five petals arranged in the following manner: two lateral petals form *wings*, a broad upper petal a *banner*, while the two lower petals are united into a *keel*. The ten stamens are united into one or two groups. The fruit, a pod such as seen on peas or beans, is called a *legume*, and is characteristic of many representatives. The pea family is one of the largest and most extensive plant families, ranking close to the grasses and orchids in numbers of species. Economically it is the most important of the dicotyledonae, providing food, fodder, dyes, gum, resins, oils, and many ornamentals. Food products include peas, beans, lentils, peanuts, yams, and soybeans. Fodder and forage plants include clover, alfalfa, and sweet clover. Sweet peas, redbud, wattles, and senna are examples of the ornamentals.

This section is divided into two parts:

    I. LEAVES PALMATELY COMPOUND, OR PINNATELY COMPOUND WITH ONLY 3 LEAFLETS. p. 145.
    II. LEAFLETS PINNATELY COMPOUND, WITH 5 OR MORE LEAFLETS. p. 149.

I. LEAVES PALMATELY COMPOUND, OR PINNATELY COMPOUND WITH ONLY 3 LEAFLETS.

1. Leaflets 4 or more. Flowers blue to purple.

LUPINE
Lupinus spp.
Latin: lupus, wolf

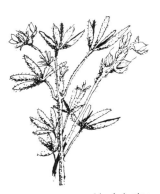

flower

Leaflets 5 to 9. Flowers blue or purple, rarely pinkish purple. The lupines are one of the largest genera within the pea family. They are called *lupus* because they rob the soil of fertility. Many species are poisonous to livestock. *L. albus* and *L. tremis* have been cultivated as ornamentals since ancient times. These were the "sad lupines" mentioned by the poet Virgil. Common species found in the area include:

fruit

KING'S LUPINE *L. kingii*. Stems growing to 25 cm, with long, silky hairs. Leaflets long-oval. Flowers arranged along the stem; flower clusters shorter than the leaves. Petals purple to blue, to 1 cm long. Fruit oval to somewhat diamond-shaped. Found in the pinyon-juniper woodland and along roadsides.

king's lupine

fruit

TALL LUPINE *L. caudatus*. Stems growing to 3 ft. Leaflets 7 to 9, long and narrow, wider at the tip than the base; both sides covered with silky hairs. Flowers along the stem in clusters (racemes) to 25 cm long. Petals dark blue, to 1 cm long. Fruit to 3 cm long. Found in the pinyon-juniper woodland and along roadsides.

1. Leaflets 3. Flowers yellow, white, or purple.

    2. Plant upright.

        3. Flowers yellow.

            4. Flowers large, to 1.5 cm. Plant of woods and canyons.

                Perennial, generally growing in clumps. Stems to 2 ft tall, smooth or with short hairs. Leaflets oval to oblong to elliptical. Flowers yellow, with 10 stamens, not united. Early spring bloomer. In canyons and open woods of the ponderosa pine and mixed conifer forests.

                BIG GOLDEN-PEA
                Thermopsis pinetorum
                Greek: *thermos*, lupine
                -opsis, resembling
                Latin: *pinetorum*, of the pines

            4. Flowers numerous and small (4 to 6 mm) in long clusters (racemes) at ends of the stems, drooping at the tip.

                Annuals or biennials with tall, smooth to slightly hairy stems, growing to 6 ft. Leaflets oblong to oval, with small teeth along margins at base of the leaflet. Flowers yellow, to 6 mm. Fruit oval, to 3.5 mm long. The plant contains coumarin (a vanilla substitute) and melilotic acid. Dried leaves have been used pharmacologically. It is a preferred pasture plant at high elevations because it reseeds readily and withstands heavy grazing. Found along roadsides, often in dense patches.

                YELLOW SWEET CLOVER
                Melilotus officinalis
                Greek: *meli*, honey; *lotos*, lotus
                Latin: *officina*, workshop

lupine

big golden-pea

yellow sweet clover

3. Flowers white or purple.

pea

   5. Flowers white.

> Annuals or biennials with tall, smooth to slightly hairy stems, growing to 6 ft tall. Leaflets oval with small teeth along the margins at the base of the leaflet. Flowers white, small, to 6 mm long, in long clusters at the end of the stem. The species is native to regions of Europe and Asia. It was first collected in the United States in 1739 and has been a major crop plant since 1900. Since antiquity the plant, sometimes called BEECLOVER or HONEYCLOVER, has been a honey plant. It has escaped cultivation; commonly found in dense patches along roadsides.
>
> WHITE SWEET CLOVER
> *Melilotus albus*
> Latin: *albus*, white

   5. Flowers purple.

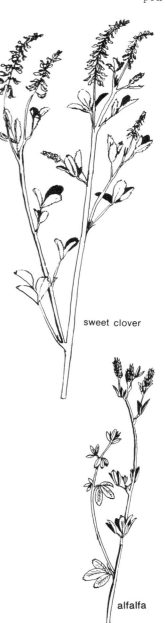

> Stems to 3 1/2 ft. Leaflets 3, oblong. Margins toothed at the tip. Flowers violet or purple, 1 cm long. Fruit coiled into a spiral. One of the most valuable forage plants, comparable to cultivated timothy as hay. Supposedly the plant originated in ancient Media, present-day, northwest Iran. Found along roadsides, trails, and in disturbed soils.
>
> ALFALFA
> *Medicago sativa*
> Greek: *medike poa*, Medic grass
> (a plant brought there from Media
> --present-day N. W. Iran)
> Latin: *sativa*, sown

2. Plant low and spreading.

   6. Flowers small, yellow tinged with orange or red.

> Annual or perennial. Stems with stiff, short hairs. Leaflets pinnately compound but appearing palmately compound. Leaflets long and narrow. Flowers yellow tinged with

147

orange, to 1 cm, solitary in leaf axils. Fruit linear. This plant is browsed by deer and other herbivores. A deep taproot makes the species very drought resistant. Found in canyons and on mesa tops in the ponderosa pine forest.

DEERVETCH
Lotus wrightii
Greek: *lotos*, a lotus plant
Honors Charles Wright (1811-1885)
American botanist

deer vetch

6. Flowers white or deep pink to rose-purple, in egg-shaped heads.

CLOVER
Trifolium spp.
Latin: *tri*, three; *folium*, leaf

Low-spreading plant with 3 leaflets. Flowers in tight heads with or without bracts at the base. CLOVERS or TREFOILS have been cultivated for pasturage since the 16th century. The plant increases the nitrogen content of the soil; the roots have nodules containing nitrogen-fixing bacteria. Usually found in disturbed soils. A number of species of clover are found in the area; they include:

RED CLOVER *T. pratense*. Stems erect or spreading, to 1 1/2 ft, sometimes hairy. Leaflets oval, smooth to toothed on margins, sparsely hairy on surfaces, spotted near the middle. Flower heads to 3 cm across; two leaf-like bracts beneath the head. Clover has had more impact upon agriculture than the potato. It has had considerable influence on European civilization by augmenting livestock production and thus food supplies. Bumblebees and some butterflies are the only insects which have probosci long enough to reach nectar within the blossoms. Dried clover flowers were once used for the treatment of whooping cough and ulcers. Found along roadsides, in arroyos, and on disturbed soils.

red clover

WHITE CLOVER *T. repens*. Stems creeping, spreading to 1 ft. Leaflets somewhat oval, often with a white blotch near the base. Leaf margins toothed. Flower head to 2.5 cm. The plant is nutritious and withstands trampling; its qualities as a honey plant make it valuable to beekeepers. The dried flowers and seeds were ground into flour for bread during the Irish potato famine. The common clover of lawns, gardens, trails, and roadsides.

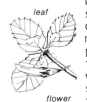
leaf
flower

white clover

II. LEAFLETS PINNATELY COMPOUND WITH MORE THAN 5 LEAFLETS.

1. Leaves with a tendril or bristle.

    2. Trailing plant, tendrils bristle-like, never climbing.

pea

Perennials with erect or trailing stems. Leaflets linear to elliptical, smooth to slightly hairy, tendrils bristle-like (the tendril an extension of the end of the leaf, replacing a leaflet). Flowers white, turning tan to yellow with age. Found at high altitudes in aspen groves and in canyons of the ponderosa pine and mixed conifer forests.

ARIZONA PEAVINE
Lathyrus arizonicus
Greek: *lathyros*, a name bestowed by Theophrastus on some leguminous plant

Arizona peavine

2. Climbing plant clinging to other vegetation with tendrils.

tendril

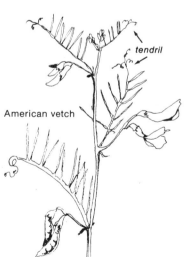

Perennial. Stems smooth to slightly hairy, trailing or climbing. Leaflets linear to oval, 6 to 14. Flowers light to dark purple, in loose clusters at the end of the stem, 3 to 10 flowers per cluster. Found in canyons of the ponderosa pine and mixed conifer forests.

AMERICAN VETCH
Vicia americana
Latin: *vicia*, vetch

tendril

American vetch

1. Leaves without tendrils.

    3. Flowers in dense spikes.

PRAIRIE CLOVER
Petalostemum spp.
Greek: *petalon*, leaf; *stamen*, thread

Stems to 2 ft tall. Leaflets 7 to 9, narrow, dotted with glands on the undersurface. Flowers in long, dense, oblong spikes. Found along roadsides in the pinyon-juniper woodland. Species commonly occurring in the area include:

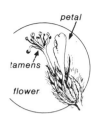

PURPLE PRAIRIE CLOVER *P. purpureum*. Stems smooth to silky hairy. Leaflets narrow, folded, smooth to silky hairy, dotted with glands on the undersurface. Flowers rose-purple with stamens extending beyond petals; flowers striking because of the yellow pollen sacs.

WHITE PRAIRIE CLOVER *P. candidum*. Stems erect or lying on the ground. Flowers white, in dense, oblong spikes.

3. Flowers not in dense spikes.

    4. Flowers greenish white, seed pods prickly.

Stems to 3 1/2 ft, leafy, dotted with minute glands. Leaflets 11 to 19, lance-shaped or oblong. Flowers to 1.3 cm long. Fruit prickly, resembling a cocklebur. Found along roadsides, in sandy areas near the Rio Grande, and in disturbed soil.

WILD LICORICE
Glycyrrhiza lepidota
Greek: *glykys*, sweet; *rhiza*, root
lepis, scale

    4. Flowers white or purple. Seed pods not prickly.

MILKVETCH, LOCOWEED
Astragalus spp.
Ancient Greek name for
some leguminous plant

Taxonomy of this genus is very complex. Only a few species are mentioned here. To identify the many species of the area, a technical botanical text such as *A Flora of New Mexico* by Martin and Hutchins is a must. Most identification schemes are based on the mature seed pods. Flowers are white, blue, pink, or purple; the keel rounded, sepals fused, with 5 teeth.

MISSOURI MILKVETCH *A. missouriensis*. Stems lying on the ground, to 15 cm long. Stems and leaves with branched hairs. Leaflets 9 to 17, elliptical. Flowers pinkish purple, banner 2 cm long. Pods to 2.5 cm long. Early spring bloomer. Found in dry canyons and along roadsides of the pinyon-juniper woodland.

STINKING MILKVETCH *A. praelongus*. Stems erect, to 1 1/2 ft tall. Leaflets 11 to 17, oval. Flowers yellow-white, banner to 2.3 mm long. Pods to 3.5 cm long.

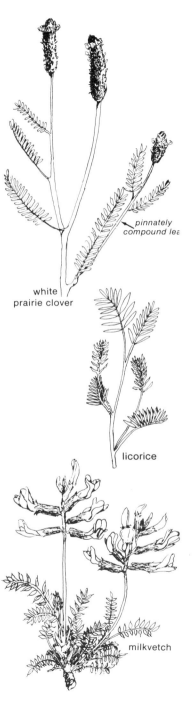

Flax Family LINACEAE

Flowers of the flax family are in fives. The delicate yellow or blue petals are fragile and fall off easily; they often wither in the hot sun. Rigid stems hold the plant erect. Leaves are narrow and alternate on the stem. Economically, the family is important for flax (*Linum usitatissimum*), widely cultivated for the fiber from which linen is made, and for its seeds, the source of linseed oil.

I. FLOWERS BLUE.

Stems to 1 1/2 ft. Leaves long, narrow, sharp-pointed. Flowers blue, to 2 cm across, easily falling as a unit. This species was an important fiber plant to some western Indians. It has been domesticated and is frequently grown in wildflower gardens. Found growing wild in disturbed soils.

WESTERN BLUE FLAX
Linum lewisii
Latin: *linum*, flax
Honors Meriwether Lewis (1774-1809)
American explorer

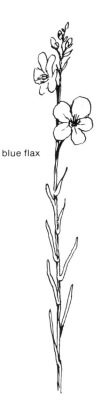

blue flax

II. FLOWERS YELLOW.

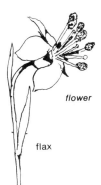

flower

flax

Stems to 1 1/2 ft. Leaves long and narrow. Flowers yellow, to 1.5 cm long, sepals lance-shaped. Found in pinyon-juniper woodland.

NEW MEXICO YELLOW FLAX
Linum neomexicana

Blazing Star Family LOASACEAE

Hooked hairs on the stems and leaves of the blazing stars are responsible for the common name STICKLEAF. The yellow to cream-colored flowers have five sepals, five petals, and many stamens. The outer filaments of the stamens are flattened and resemble petals, giving the flower the appearance of having many petals. Stems are often conspicuously white. Leaves are sometimes pinnatifid and alternate on the stem. Flowers open in the evening and are pollinated by night-flying insects. Legend claims that Indians attached the sticky leaves to the legs of a first-time horseback rider to give him a good grip.

I. FLOWERS 1 - 3 CM WIDE.

Stems to 1 ft tall, lustrous white, smooth to slightly hairy. Leaves pinnatifid, toothed or smooth. Flowers to 1.5 cm long, light yellow, blooming in subdued light. Found along roadsides and trails of the pinyon-juniper woodland and ponderosa pine forest.

STICKLEAF
Mentzelia pumila
Honors C. Mentzel, 17th-century
German botanist
Latin: *pumilus*, dwarf

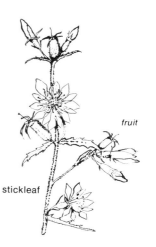

fruit

stickleaf

II. FLOWERS 2 - 4 MM WIDE.

Stems to 1 ft, white to greenish white. Leaves linear to lance-shaped. Lower leaves with smooth margins; upper pinnatifid. Flowers yellow, in leaf axils. Indians parched the oily seeds of the WHITE-STEMMED BLAZING STAR and ground them into meal. Found in disturbed soils.

WHITE-STEMMED BLAZING STAR
Mentzelia albicaulis
Latin: *albus*, white
Greek: *kaulos*, stem

Mallow Family MALVACEAE

Plants of the mallow family are characterized by the fusion of the stamens into a tube. This tube extends beyond the five sepals and petals. Flowers may be either red, orange or white and rolled in the bud. The stems and leaves are often covered with star-shaped hairs. Leaves are simple, alternate and may be lobed, toothed, or dissected. The hollyhock, rose-of-Sharon, okra, cotton, and hibiscus are all domesticated representatives of this family.

I. **FLOWERS BRICK RED OR ORANGE. PLANTS ERECT. PLANTS OF DRY SOIL.**

globe mallow

GLOBE MALLOW
Sphaeralcea spp.
Greek: *sphaira*, sphere; *alkea*, mallow

This is a complex genus whose species are almost impossible for an amateur to identify with certainty. Identification at the species level depends on precise characterization of the seeds and fruits. Representative species found in the area include:

flower

GLOBE MALLOW *S. incana*. Stems growing to 2 ft. Leaves 3-lobed. Flowers light orange. Found in pinyon-juniper woodland and ponderosa pine forest.

SCARLET GLOBE MALLOW *S. coccinea*. Stems to 2 ft. Leaves deeply divided. Flowers red; lacks the 3 bracts usually found beneath mallow flowers. Called *yerba de la negrita* by the Spanish. Crushed leaves mixed with salt were used as a poultice for mosquito and ant bites. Found in pinyon-juniper woodland and ponderosa pine forest.

II. **FLOWERS WHITE, PLANTS TALL, OF MOIST PLACES.**

fruit

An erect plant growing up to 3 ft. Leaves rounded, palmately lobed, smooth or with a few hairs on the surfaces; lower leaves shallowly lobed. Leafstalks long. Flowers white, rarely purple, to 2 cm long; in long clusters. Found along streams in ponderosa pine and mixed conifer forests.

white prairie mallow

WHITE PRAIRIE MALLOW
Sidalcea candida
Latin: *sida*, pomegranate or mallow;
candidus, white

III. FLOWERS WHITE. PLANTS LOW, SPREADING; ON DISTURBED GROUND.

CHEESEWEED
Malva spp.
Latin: *malva*, mallow

flower

Low, bushy annuals growing to 1 ft across. Leaves large, almost circular, often with a red spot at the base, with 5 to 7 shallow lobes. Flowers bluish to pinkish, in clusters at the base of the leafstalk. The fruit resembles a round of cheese, giving the common name CHEESEWEED. These buttons are edible when soft. Found in moist soils of lawns, roadsides, trails and along streets and sidewalks. Species found in the area include:

cheeseweed

fruit

COMMON CHEESEWEED *M. neglecta*. Flowers white to blue, to 1 cm.

CHEESEWEED *M. parviflora*. Flowers pink to lilac, to 5 mm.

IV. FLOWERS PURPLE. OF DISTURBED OR MOIST GROUND.

fruit

Stems to 1 ft, hairy or smooth. Leaves somewhat arrow-head shaped. Flowers purple, 2 to 2.5 cm long. Fruits with conspicuous awns on the back. Found in disturbed soils.

ANODA
Anoda cristata
Anoda: Ceylonese name for a similar plant
Latin: *crista*, crest

anoda

Four-O'clock Family NYCTAGINACEAE

Characteristically, plants of the four-o'clock family have showy, colored sepals and no petals. Below the sepals is a leafy bract which is sometimes enlarged and papery. The sepals may be bell-shaped, dish-shaped, or funnel-like, usually three- to five-lobed. Stamens may be one to many, but usually number three to five. Our species have purple or rose-purple flowers. Except for ornamentals such as four-o'clock, sand verbena and bougainvillea, the family is of little economic importance. Species commonly found in the area include:

I. LARGE SHOWY PLANTS SPREADING TO 2 FT; FLOWERS PURPLISH RED.

flower   fruit

Large showy plant with stems spreading to 2 ft. Leaves oval, softly hairy, on long leaf stems. Flowers purplish red, often 5 cm long; usually 3 to 6 flowers within 1 bract. The root was supposedly used by the Indians to induce visions. Found under pinyons and junipers in the pinyon-juniper woodland.

SHOWY FOUR-O'CLOCK
Mirabilis multiflora
Latin: *mirabilis*, wonderful
multus, many; *flos*, flower

showy four-o'clock

II. TALL, MUCH—BRANCHED PLANT; FLOWERS PINK TO PURPLISH RED.

flower

bracts

Much-branched perennial growing to 2 ft. Stems sticky-hairy. Leaves linear to lance-shaped; margins slightly toothed. Flowers pink to purplish red, to 1 cm long, sticky-hairy; bracts with clusters of 3 flowers. Found in dry soils of the pinyon-juniper woodland and ponderosa pine forest.

DESERT FOUR-O'CLOCK
Oxybaphus linearis
Greek: *oxybaphos*, a shallow dish
Latin: *linea*, line

III. TRAILING PLANT.

See VINES AND TRAILING PLANTS, p. 14.

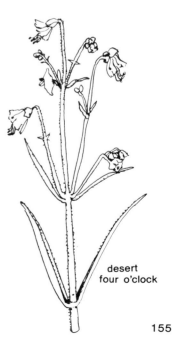
desert four o'clock

Evening-Primrose Family ONAGRACEAE

Flower parts in the evening-primrose family are in fours or multiples of four. There are four petals and sepals, eight stamens. The long tube below the petals is called a *hypanthium*. Because of its length it can be confused with a flower stalk. The ovary is within a swelling at the base of the hypanthium. Leaves may be opposite or alternate. Margins are smooth to pinnatifid. Many members of the genus *Oenothera* open in the evening and wither in strong sunlight, giving rise to the common name EVENING-PRIMROSE.

This section is divided into three sub-sections, according to flower color.

I. FLOWERS PINK TO SCARLET.

GAURA
Gaura spp.
Greek: *gauros*, superb

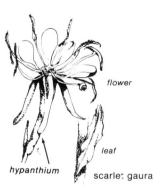

scarlet gaura

Weedy plant growing from 1 to 6 ft. Petals small, mostly on one side, making the flower somewhat asymmetrical. Flowers in spikes or loose inflorescences along the stem. Species found in the area include:

SCARLET GAURA *G. coccinea*. Grows to about 1 ft tall. Stems smooth to roughly hairy. Leaves lance-shaped to oblong, the upper ones lance-shaped; margins wavy-toothed. Flowers to 6 mm long, pink to whitish, becoming red. Stamens either yellow or red. Found in the pinyon-juniper woodland and ponderosa pine forest.

habit

TALL GAURA *G. parviflora*. Grows to 6 ft. Leaves lance-shaped. Flowers pink, 2 mm long, in spikes that nod at the tip. Stamens with rose-colored stalks and reddish anthers. Found along roadsides in the pinyon-juniper woodland and ponderosa pine forest.

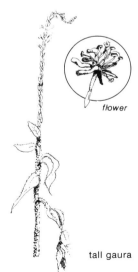

tall gaura

II. FLOWERS PURPLE.

WILLOWWEED, FIREWEED
Epilobium spp.
Greek: *epi*, upon; *lobos*, lobe

Stems from 15 cm to 3 ft. Leaves opposite. Flowers purple, petals notched. Representative species of the area are:

156

FIREWEED *Epilobium angustifolium*. A showy species growing to 3 ft. Leaves lance-shaped, lateral veins curved into hooks along margins. Flowers 2 cm long; petals narrowed at the base. Stamens 8, purple. Seeds with cottony hairs that catch in the wind, assuring wide seed distribution. Growing in patches in recently logged or burned areas, and along roadsides.

WILLOWWEED *Epilobium ciliatum*. May grow to 3 ft, but generally less than 1 ft. Leaves oval to lance-shaped. Flowers purple, to 6 mm long; sepals 2, oval to lance-shaped.

III. FLOWERS WHITE OR YELLOW.

1. Flowers yellow.

    2. Plant to 3 ft tall.

        A tall plant, growing to 3 ft. Stems covered with stiff, red hairs. Leaves lance-shaped. Sepals 2 to 2.5 cm, reddish, with short, sticky hairs. Petals yellow, turning reddish pink with age. Found along roadsides, trails, and disturbed soils of the pinyon-juniper woodland, ponderosa pine and mixed conifer forests.

        HOOKER'S EVENING-PRIMROSE
        Oenothera hookeri
        Greek: *oinotheras*, a plant whose root
        smells like wine (Gr. *oinos*, wine)
        Honors William Hooker (1785-1865)
        curator of Kew Botanical Gardens

    2. Plant growing to 1 1/2 ft tall.

        A low, spreading plant no more than 1 1/2 ft tall. Sepals yellowish, blotched with red, 10 to 15 cm long. Petals yellow, turning red on drying.

        SUNDROPS
        Calylophus hartwegii
        Greek: *kalyx*, cup; *lophos*, crest
        Honors Carl Hartweg (1812-1871)
        German botanical explorer

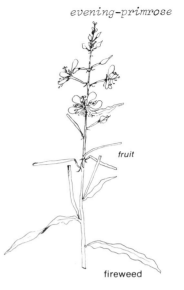

evening-primrose

fruit

fireweed

Hooker's evening-primrose

sundrops

1. Flowers white.

EVENING-PRIMROSE
Oenothera spp.

The evening-primroses may open in the morning, evening, or during the day. Generally the flowers remain open for only a day, then wither. Several of the evening-blooming species open abruptly at dusk and are pollinated by such night-flying insects as the sphinx moth. There are a number of evening primrose species and their identification is usually laborious. Most are found in disturbed soils. Three common species found in the area are:

white stemless evening-primrose

WHITE STEMLESS EVENING — PRIMROSE *O. caespitosa*. Stemless. Leaves with tuft of flowers among them. Petals white, heart-shaped. Hypanthium tinged with red, becoming pink with age; 2.5 to 4 cm long.

CUTLEAF EVENING—PRIMROSE *O. coronopifolia*. Grows to 25 cm tall. Leaves fused with segments arranged like teeth in a comb. Flowers white; sepal tips fused.

PRAIRIE EVENING — PRIMROSE *O. albicaulis*. Grows to 1 1/2 ft. Leaves divided into narrow lobes. Flowers white with conspicuous hairs in the throat. Sepal tips directed backward.

prairie evening-primrose

Oxalis Family OXALIDACEAE

Leaflets of oxalis resemble those of clover or shamrock. The violet-colored flowers have five sepals, five petals, and ten stamens, five of which are long and five short. Bases of the stamens are all joined into a tube. At night or in diffused light the compound leaves fold back and the delicate purple flowers close. The watery juice of the plant is sour, containing oxalic acid. Plants grow from bulb-like tubers, and multiply rapidly. In gardens they can become noxious weeds.

violet woodsorrel

Stemless plant growing to 20 cm. Leaves half as long as flowering stalks; leaflets to 2 cm. Flowers purple, to 2 cm long. Moist canyons in ponderosa pine and mixed conifer forests; open meadows.

VIOLET WOODSORREL
Oxalis violacea
Greek: *oxalis*, acid
Latin: *viola*, violet

Plantain Family PLANTAGINACEAE

Plants of the plantain family have small, inconspicuous flowers arranged in dense spikes. Sepals and petals are four; the petals are papery. Leaves are inconspicuously veined. The seeds of flaxseed plantain *P. psyllium*, a native of Europe and Africa, have been grown and sold for medicinal use. Species commonly found in the area include:

I. PLANTS TO 15 CM TALL, LEAVES WOOLLY OR SILKY HAIRY.

Small weedy plant growing to 15 cm. Leaves basal. Flowers inconspicuous, 4-lobed, 2 mm long; in a dense, woolly spike. Indian-wheat is highly palatable to livestock and an important range plant. Found in dry soils of the pinyon-juniper woodland.

WOOLLY INDIAN-WHEAT
Plantago purshii
Latin: *plantago*, ancient
name for this family
Honors Frederich Pursh (1774-1820)
German-born botanist and curator

flower

Indian-wheat

II. PLANTS TO 1½ FT, LEAVES SMOOTH, CONSPICUOUSLY 5 TO 7 VEINED.

Perennial growing to 1 1/2 ft. Leaves large, to 15 cm long, oval to broadly egg-shaped. Veins conspicuous, appearing parallel. Margins wavy or toothed. Flowers to 1 mm long, in a spike sometimes growing to 1 ft. A naturalized weed sometimes used by Spanish New Mexicans as a headache remedy. Found in disturbed soils of lawns, roadsides, and trails.

flower
fruit

RIPPLESEED PLANTAIN
Plantago major
Latin: *magnus*, great

rippleseed plantain

Phlox Family POLEMONIACEAE

Plants of the phlox family have tubular or salver-like flowers which are twisted in the bud. The flowers have five sepals, five petals, five stamens, and three stigmas. Economically the family is important for a few ornamental species such as phlox, gilia, and polemonium.

I. FLOWERS DARK PURPLE, BELL-SHAPED WITH SKUNK-LIKE SMELL

An odoriferous perennial growing to 3 ft, covered with glandular hairs. Leaves pinnately compound with leaflets lance-shaped to oblong. Flowers purple, in a more or less flat-topped cluster. Found in meadows at the higher altitudes.

JACOB'S LADDER
Polemonium foliosissimum
Greek: *polemos*, war
Latin: *folium*, leaf, *-issimus*, very many

Jacob's ladder

II. FLOWERS LIGHT BLUE, SALVER-LIKE.

Stems to 1 ft, smooth to sticky, much branched. Leaves pinnatifid, divided into fine divisions. Flowers to 5 cm long, in more or less flat-topped clusters. These flowers attract night-flying moths that feed on the nectar. Blooming in autumn. Found along roadsides.

PALE TRUMPET
Ipomopsis longiflora
Greek: *ips*, worm: *-opsis*, resembling
Latin: *longus*, long; *flos*, flower

pale trumpet

III. FLOWERS SCARLET, FUNNEL-LIKE.

Stems to 2 ft. A biennial, growing the first year as a rosette of leaves, the second year as a tall, erect stem bearing pinnately divided leaves. Flowers red, more or less mottled with yellow, stamens extending beyond petals. This plant reportedly was used medicinally and ceremonially by some southwestern tribes. The bright red color attracts hummingbirds, and it is often called "hummingbird flower."

DESERT TRUMPET, SKYROCKET
Ipomopsis aggregata
Latin: *aggregatus*, added to
from *gregis*, herd

skyrocket

Buckwheat Family POLYGONACEAE

Flowers of the buckwheat family are small and lack petals. Sepals are papery, pink to white. Many species have basal leaves, and also have alternate or opposite leaves. Fruit is three-sided with a hard shell. Economically the family is important for food plants such as buckwheat and rhubarb, and for ornamentals including silver-lace vine, sea-grape, and mountain-rose vine. Many other representatives are noxious weeds. The genus *Rumex* is discussed under WEEDS, p. 109.

I. FLOWERS PINK, STEMS PROSTRATE. PLANTS OF DAMP SOIL.

flowers

A prostrate perennial, stems to 1 ft. Leaves lance-shaped, bearing a hinge-like joint near the sheath which surrounds the stem. Flowers to 3.5 mm, in clusters in the axils of the leaves. Found near the Rio Grande or along streams.

SMARTWEED
Polygonum aviculare
Greek: *polys*, many; *gony*, knee
Latin: *aviculare*, bird-like

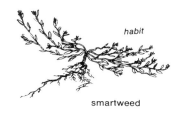
habit

smartweed

II. FLOWERS PINK TO WHITE, STEMS UPRIGHT, LEAVES MOSTLY AT THE BASE OF THE PLANT. PLANTS OF DRY SOIL.

WILD BUCKWHEAT
Eriogonum spp.
Greek: *erion*, wool; *gony*, knee

A large genus. Leaves usually basal, smooth or covered with whitish hairs. Individual flowers on tiny stalklets. Petals absent, sepals white to pinkish. Flowers in heads, in umbrella-shaped clusters, or clustered along the stem. Found in the pinyon-juniper woodland or disturbed soils.

Some species found in the area include:

E. alatum Grows to 3 ft. Flowers in elongate clusters, yellowish.

E. racemosum Grows to 2 1/2 ft. Flowers in a spike-like inflorescence, rose-colored.

E. cernuum Grows to 1 ft. Flowers in umbrella-shaped clusters, white to rose.

E. abertianum Grows to 2 ft. Flowers in a leafy inflorescence, white or yellow, tinged with pink or red.

E. racemosum

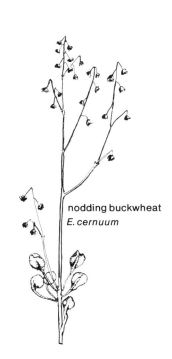
nodding buckwheat
E. cernuum

III. FLOWERS WHITE, STEMS WOODY. PLANTS OF ROCKY CANYONS.

flower

Stems to 1 ft. Leaves spatula-like, tapering at the base. Leafstalk nearly as long as the leaf blade. The upper leaf surfaces smooth, lower hairy. Flowers creamy white, silky-hairy, in flat-topped clusters.

ANTELOPE SAGE
Eriogonum jamesii
Honors Edwin James (1797-1861)
American physician and botanist

antelope sage

Woody perennial. Stems growing to 1 ft. Leaves linear, 2-4 cm long; upper surface smooth, lower surface hairy. Margins smooth, inrolled. Flowers white to pink, 2.5 to 3.5 mm long, white to pink, midveins pink. Dry mesas.

WILD BUCKWHEAT
Eriogonum leptophyllum
Greek: *leptos*, thin; *phyllum*, leaf

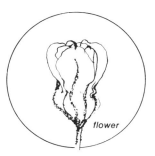
flower

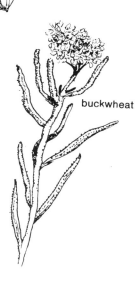
buckwheat

Primrose Family PRIMULACEAE

Flowers of the primrose family have fused petals, radial symmetry, and flower parts in fives. Only two genera of this family have been reported for the area, ROCK JASMINE *Androsace* and SHOOTING STAR *Dodecatheon*. Leaves are opposite, basal or in circles around the stem. The family is economically important for a few ornamentals, including cyclamen.

I. FLOWERS WHITE, IN UMBRELLA-LIKE CLUSTERS.

A delicate plant growing to 10 cm, blooming early in the spring and throughout the summer. Leaves basal. Flowers 5-lobed, bell-shaped, appearing like tiny stars. Found in woods and canyons of mixed conifer and ponderosa pine forests.

flower

ROCK-JASMINE
Androsace septentrionalis
Greek: *andros*, polyp
Latin: *septem*, seven; *triones*, oxen
(refers to the seven stars of
the constellation Ursa Major)

rock--jasmine

shooting star

II. FLOWERS PINK, 2 TO 20 FLOWERS ON A LEAFLESS STEM. PETALS DI-RECTED BACKWARD, ANTHERS FORMING A POINT.

Stems to 1 1/2 ft. Leaves narrowly lance-shaped to elliptical, in a rosette at the base of the plant. Margins smooth to wavy-toothed. Flowers purple-rose with a dark, wavy line at the base. Anthers forming a point which projects beyond the petals. Found along streams or in wet meadows of higher elevations.

SHOOTING STAR
Dodecatheon pulchellum
Greek: *dodeka-*, twelve; *theos*, gods
Latin: *pulcher*, beautiful

Buttercup Family RANUNCULACEAE

Many members of the buttercup family do not appear to be related. Except for MONKSHOOD and LARKSPUR, the flowers are symmetrical. Some members, like COLUMBINE, have nectar-containing structures called spurs. In other genera the sepals are petal-like (CLEMATIS, WINDFLOWER). Commonly there are five sepals and five petals, but there may be as many as twenty-three or as few as three. Stamens are usually numerous (20 to 50), as are the pistils (5 to 300). Leaves are often divided or compound. Generally members of the family are found in cool, moist sites at higher elevations. Economically the family is important because it contains a large number of ornamentals, such as anemone, delphinium, peony, hellebore, and ranunculus.

This section is divided into five sub-sections:

    I. FLOWERS ASYMMETRICAL, BLUE OR WHITE, p. 165
   II. FLOWERS SYMMETRICAL BUT SPURRED, RED OR BLUE, p. 166
  III. FLOWERS SYMMETRICAL, PETALS YELLOW OR WHITE, STAMENS ON A HUMP IN THE CENTER, p. 166
  IV. FLOWERS SYMMETRICAL, SEPALS PETAL-LIKE, LAVENDER OR WHITE, p. 167
   V. FLOWERS GREEN, INCONSPICUOUS, LEAVES APPEARING FERN-LIKE, p. 168

I. FLOWERS ASYMMETRICAL, BLUE OR WHITE.

1. Sepals shaped like a hood.

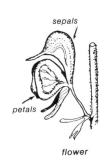
flower

monkshood

Tall, distinctive plant growing 6 ft tall. Leaves palmately 3- to 5-lobed. Flowers in loose clusters at the top of the stem. Sepals and petals blue, rarely white. Top sepals hooded; the other sepals and petals inside the hood. All parts of this plant are poisonous, containing the alkaloids aconitine and aconine. A heart and nerve sedative, aconite, is obtained from this plant. Along streams in the mixed conifer forest.

                                 MONKSHOOD
                        Aconitum columbianum
                      Greek: *akon*, a dart
columbiana, North American (after Columbus)

1. Upper sepals and petals forming a spur.

LARKSPUR
Delphinium spp.
Latin: *delphinus*, dolphin

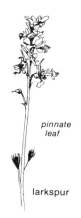

pinnate leaf

larkspur

Species found in the area include:

spur

flower

WHITE LARKSPUR *D. virescens*. Grows to 1 ft. Leaves pinnatifid with curling hairs on the surface. Flowers bluish white, the lower petals notched. Found in dry, rocky soils near the Rio Grande.

TALL PURPLE LARKSPUR *D. occidentale*. Grows to 6 ft, stems hollow. Leaves pinnatifid. Flowers dark purple. Found in meadows of high altitudes.

II. FLOWERS SYMMETRICAL BUT SPURRED, RED OR BLUE.

COLUMBINE
Aquilegia spp.
Latin: *aquila*, eagle

spur

little red columbine

Leaves usually divided into several lobes. Flowers distinctive; petals with long spurs directed backwards. Sepals not spurred. Sepals and petals may be of the same or different colors. Species found in the area include:

LITTLE RED COLUMBINE *A. triternata* or *A. elegantula*. Erect or sprawling plant growing to 1 ft. Flowers nodding, to 3.5 cm long; petals yellow or reddish, sepals red. Found in moist canyons of the ponderosa pine and mixed conifer forests.

Colorado blue columbine

COLORADO BLUE COLUMBINE *A. caerulea*. Grows to 2 ft, stems sticky-hairy. Leaves divided into 3 parts, with smooth or sparsely hairy surfaces. Sepals blue, to 3.5 cm long; petals white to 2.5 cm long. Spurs twice as long as the petals, blue or rarely white. Found in subalpine meadows.

III. FLOWERS SYMMETRICAL, PETALS YELLOW OR WHITE, STAMENS ON A HUMP IN THE CENTER.

hump

sepals

BUTTERCUP
Ranunculus spp.
Latin: *rana*, frog

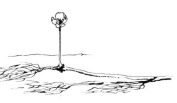

white water-crowfoot

Flowers with 5 green sepals and, usually, 5 white or yellow petals, though the number of petals can vary. Petals generally having shiny or waxy appearance. Many stamens and pistils situated on a hump. Leaves variously divided or lobed. Most species found in moist sites and along streams. Species occurring in the area include:

HEART-LEAVED BUTTERCUP *R. cardiophyllus*. Erect plant to 1 1/2 ft. Leaves heart-shaped, margins scalloped. Petals small, yellow.

MACOUN'S BUTTERCUP *R. macounii*. Stems hairy, erect or bending to the ground, sometimes rooting. Leaves triangular. Petals yellow.

WHITE WATER-CROWFOOT *R. aquatilis*. Stems submerged in water. Leaves finely dissected. Flowers white or tinged with yellow at base, projecting above surface of water.

HOMELY BUTTERCUP *R. inamoenus*. Perennial growing to 1 ft. Leaves at the base with long leafstalks, blades rounded, 3-lobed. Upper leaves 3 times divided. Petals to 5 mm long.

buttercup

heart-leaved buttercup

flower

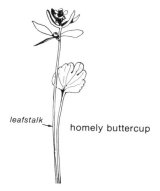

leafstalk — homely buttercup

IV. FLOWERS SYMMETRICAL, SEPALS PETAL-LIKE, LAVENDER OR WHITE.

1. Sepals bluish or lavender, plant to 15 cm.

Flowering stalk appearing before the leaves; a circle of narrow bracts below the flower. Plant covered with long silky hairs. Leaves 3 times divided. The species contains a poisonous drug, anemonin, which has caused livestock losses. Indians crushed the leaves for treatment of rheumatism. Leaves were applied as a poultice, but if left long in contact with the skin caused blistering. Pasque flower is the state flower of South Dakota. One of the earliest spring bloomers. Found in canyons and on slopes of the ponderosa pine forest.

seed head

leaf

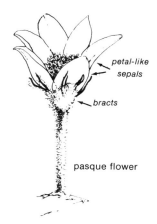

petal-like sepals

bracts

pasque flower

PASQUE FLOWER
Pulsatilla ludoviciana
Latin: *pulsare*, to beat
ludoviciana, of Louisiana

1. Sepals greenish white.

167

Leaves both at the base and along the stem, palmately parted. Upper leaves opposite or in a circle around the stem. Sepals petal-like, greenish white. Fruiting heads elongated, candle-shaped. Found in damp soil along stream banks in mixed conifer forest.

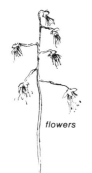

CANDLE ANEMONE
Anemone cylindrica
Greek: *anemos*, wind; *kylindros*, cylinder

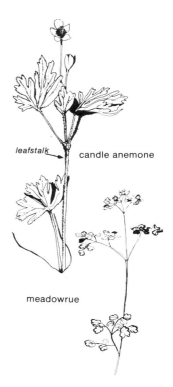

V. FLOWERS GREEN, INCONSPICUOUS, LEAVES APPEARING FERN-LIKE.

Delicate plant growing to 2 ft. Leaves 3 times compound, each leaflet round and 3-lobed. Male and female flowers on separate plants. Male flowers appearing like little tassels; female flowers inconspicuous clusters. Canyons and woods of the ponderosa pine and mixed conifer forests.

FENDLER MEADOWRUE
Thalictrum fendleri
Greek: *thallein*, to bloom
Honors August Fendler (1813-1883)
German born naturalist

Rose Family ROSACEAE

Many members of this family are shrubs or trees. CINQUEFOIL *Potentilla* and STRAWBERRIES *Fragaria* are two common herbaceous members. Plants of the rose family characteristically have five sepals, five petals, many stamens, and many pistils. Stamens sit on the edge of a floral cup. Often there are five bracts beneath the sepals (a characteristic not found in the buttercups, which are often confused with the roses). Leaves are compound. This family is of considerable economic importance. There are many fruit-producing members such as the apple, pear, quince, cherry, plum, prune, peach, nectarine, apricot, almond, blackberry, raspberry, loganberry, strawberry, and sloeberry. There are also many ornamental shrubs and trees within the family, including rose, spirea, ninebark, cotoneaster, firethorn, hawthorn, flowering quince, mountain ash, shrubby cinquefoil, and Japanese cherry.

I. FLOWERS WHITE, LEAVES WITH THREE LEAFLETS.

Stemless perennial with runners. Leaves with 3 leaflets; leafstalk with spreading hairs. Leaf surfaces silky-hairy when young, smooth with age. Margins of the leaves coarsely toothed. Flowers white, 5 additional bracts beneath the flower. Fruits to 3 cm, sweet. Relished by animals and birds, and making a delightful snack while hiking. The first records of cultivation of strawberries are from France in the 14th century. The wild Virginia strawberry found by early American colonists surpassed European varieties and became the preferred cultivar. Canyons and forests of the ponderosa pine and mixed conifer forests.

WILD STRAWBERRY
Fragaria americana
Latin: *fraga*, strawberry

II. FLOWERS YELLOW, FLORAL CUP BEARING A RING OF HOOKED PRICKLES.

Stems growing to 3 ft, sticky-hairy. Leaves pinnately compound, leaflets unequal with smaller leaflets between large ones. Surfaces hairy, especially on the veins. Margins coarsely toothed. Petals yellow, twice as long as the sepals, in spike-like clusters at the end of the stems. Found in moist canyons of the mixed conifer forest.

agrimony

AGRIMONY
Agrimonia striata
Latin: *agrimonia*, ancient name for the plant; *striatus*, grooved

III. SEPALS REDDISH-PURPLE, PETALS YELLOW TINGED WITH PURPLE. PLANTS OF HIGH-ALTITUDE MEADOWS.

Stems to 2 ft, softly hairy. Leaves pinnately compound, basal leaves larger than stem leaves. Leaflets oblong-oval, sepal lobes russet-pink to reddish purple. Petals yellow tinged with purple, hanging downward. The fruit feathery, giving plant its common name.

old-man's whiskers

OLD-MAN'S WHISKERS
Geum triflorum
Greek: *genein*, to taste
Latin: *tri-*, three; *flos*, flower

IV. FLOWERS YELLOW, 5 BRACTS ALTERNATING WITH THE SEPALS.

POTENTILLA, CINQUEFOIL
Potentilla spp.
Latin: *potens*, powerful

silverweed

A large genus of perennials. Flowers borne in loose, open clusters at ends of the stems; yellow, cup-shaped, with 5 petals; 5 sepal-like bracts alternate with the 5 sepals. Leaves pinnately or palmately compound, variously margined. The common name CINQUEFOIL refers to 5 finger-like leaves present in some species. Potentillas have been used medicinally for centuries. Some are high in tannic acid. Found in canyons, moist meadows, and forests throughout our range.

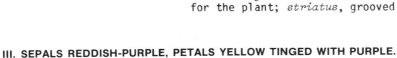
sepals / bract / flower underside

rose

SILVERWEED *P. anserina*. Matted plant producing runners. Leaves pinnately compound; 9 to 31 oblong leaflets present with smaller ones interspersed. Upper leaf surfaces silky-hairy, lower woolly-hairy. Margins toothed. This species is often found in areas where overgrazing has occurred. It can withstand trampling. Found in moist or disturbed soils.

SILVERY POTENTILLA *P. hippiana*. Leaves pinnately compound, with 7 to 13 coarsely toothed leaflets; surfaces silky or woolly-hairy. Petals yellow, 1 cm long or longer, with bracts longer than the sepals.

BEAUTY POTENTILLA *P. pulcherrima*. Grows to 2 ft tall. Leaves at the base; either pinnately or palmately compound, with 7 leaflets, sometimes 5 to 11. Upper surfaces smooth; lower hairy, whitish. Margins coarsely toothed. Flowers yellow, bracts shorter than sepals.

NORWAY POTENTILLA *P. norvejica*. Stems to 1 ft, reddish. Leaves with 3 leaflets. Flowers yellow, bracts the same length as sepals.

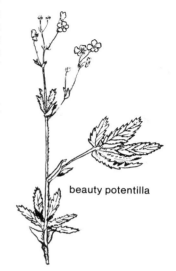

beauty potentilla

Madder Family RUBIACEAE

Plants of the madder family have square stems. Leaves are linear to lance-shaped, in circles around the stems; leaf numbers vary from species to species. Flowers are tiny, numerous, with four to five sepals, four to five petals and equal numbers of stamens. Some members of the family have nitrogen-fixing bacteria in nodules on the leaves. This family is economically important for several tropical crops, including coffee, quinine, and ipecac. A number of ornamentals including gardenia, bead plant, and partridge berry belong to this family.

I. LEAVES 4, IN A CIRCLE AROUND THE STEM.

Stems growing to 2 ft. Whole plant covered with minute barbed hairs which catch in clothing. Plant sometimes called CLEAVERS. Leaves linear to lance-shaped with 3 prominent veins; in a group of 4 around the stem. Flowers white, to 4 mm in diameter, saucer-shaped. Species increasing in overgrazed areas. Found in canyons of the ponderosa pine and mixed conifer forests.

NORTHERN BEDSTRAW
Galium boreale
Latin: *gala*, milk; *borealis*, northern

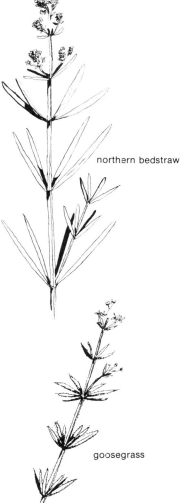

northern bedstraw

II. LEAVES 5 TO 8, IN A CIRCLE AROUND THE STEM.

Stems spreading to scrambling, growing to 2 1/2 ft. Leaves 5 to 8 in a circle around the stem, oblong lance-shaped to linear-oblong. Margins and veins with hairs; only 1 vein prominent. Flowers white to greenish, with 4 lobes, arranged in flat-topped clusters, 1 to 3 per cluster. This plant was used by early settlers to stuff mattresses, which is the origin of its common name BEDSTRAW. Seeds of this species may be roasted, ground, and used as a coffee substitute. Indians used the roots of a related species, *G. tinctorum*, to prepare a red or yellow dye. Another species found in the area is FRAGRANT BEDSTAW (*Galium triflorum*). Stems of this species have no hairs. Flower are white to yellow or greenish white. Found in canyons and woods throughout the area.

goosegrass

GOOSEGRASS
Galium aparine
Latin; *aparis*, unequal

Saxifrage Family SAXIFRAGACEAE

Most members of this family, CLIFFBUSH (p. 39), MOCK ORANGE (p. 40), CURRANT and GOOSEBERRY (p. 36), and FENDLERBUSH (p. 39) are shrubs. ALUMROOT *Heuchera* and SAXIFRAGE *Saxifraga* are the two herbaceous genera found in the area. Flowers of these species are small, in fives, and found on a long leafless stem. Basal leaves are palmately veined. Alumroots have only five stamens, while the saxifrages have ten.

I. FLOWERS GREENISH-WHITE, ARRANGED ALONG A LEAFLESS STEM.

Stems to 20 cm. Leaves all basal. Leaves 7- to 9-lobed, with hairs on the margin and rounded teeth. Surfaces sticky-hairy to smooth. Plant is called ALUMROOT because of the puckery taste of the roots. *H. americana* is the source of the drug heuchera which was used as an antiseptic and astringent. Certain Indian tribes used the powdered roots for sores. Found in moist canyons, damp woods, and rocky places throughout the area.

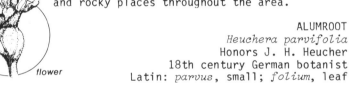
flower

ALUMROOT
Heuchera parvifolia
Honors J. H. Heucher
18th century German botanist
Latin: *parvus*, small; *folium*, leaf

basal leaf

alumroot

II. FLOWERS WHITE, IN A BALL-SHAPED CLUSTER AT TOP OF A LEAFLESS STEM.

SAXIFRAGE
Saxifraga spp.
Latin: *saxum*, stone; *fragere*, to break

Two species found in the area are:

SAXIFRAGE *S. rhomboidea*. Stems to 1 ft. Leaves at the base of the plant, oval, scalloped or toothed on the edges. Surfaces smooth. Flowering stems sticky-hairy and downy. Flowers in a head; petals white, notched. In mountain meadows.

flower

SPOTTED SAXIFRAGE *S. bronchialis*. Stems to 15 cm tall. Plant forms mats. Leaves all at the base, lance-shaped, spine-tipped; margins hairy. Flowers white dotted with yellow or reddish purple. On rocks or in rocky places of upper canyons.

spotted saxifrage

Figwort Family SCROPHULARIACEAE

Some members of the figwort family have uniquely asymmetrical flowers. Those of LITTLE RED ELEPHANT resemble an elephant's head, while MONKEYFLOWER looks like a monkey's face. Stems are round; leaves are opposite. Most flowers are asymmetrical, with five sepals, petals, and stamens. In some species only four of the stamens are fertile, the fifth being a sterile stamen, one having no anther. This family contains a number of garden ornamentals, including the snapdragon.

This section is divided into the following subsections:

    I. FLOWERS SLIGHTLY ASYMMETRICAL; FLOWERS WHEEL-LIKE, p. 174
    II. FLOWERS STRONGLY ASYMMETRICAL (2-LIPPED); PURPLE, BLUE, WHITE, OR RED, p. 175
    III. FLOWERS STRONGLY ASYMMETRICAL; YELLOW, p. 177
    IV. FLOWERS ASYMMETRICAL; COLORFUL RED, ORANGE, OR YELLOW BRACTS SURROUNDING THE FLOWER, p. 178

I. FLOWERS SLIGHTLY ASYMMETRICAL; FLOWERS WHEEL-LIKE.

1. Flowers yellow or white. Plant tall and robust. Stamens 5.

flower

Biennials growing over 6 ft. Stems stout, covered with densely branched, woolly hairs. Stem leaves up to 1 1/2 ft long, elliptical to elongate, wider near the tip than the base; margins coarsely toothed. Leaves yellow-green, surfaces covered with branched, woolly hairs. Leaves of the first year very large, in a rosette at the base of the plant. Flowers in a dense, spike-like inflorescence. Petals to 2.5 cm wide, stamens with anthers 5. This plant was introduced from the Mediterranean region when the Spanish arrived in the Southwest. It was dipped in tallow and used as a lampwick, thus the common name *candelaria*. It has also been used as a substitute for tobacco and as a medicinal plant to treat pulmonary disorders. Found commonly along roadsides in the pinyon-juniper woodland and ponderosa pine forest.

mullein

                                  MULLEIN
                           Verbascum thapsus
Latin: *verbascum*, mullein
Thapsis, an ancient town
in north Africa

1. Flowers blue. Stems creeping. Plant of wet ground. Stamens 2.

flower

fruit

Stems growing to 2 ft; erect or lying on the ground, somewhat succulent, branched or unbranched, smooth. Leaves lance-shaped to oval. Flowers blue to white, to 1 cm wide; in long clusters in the axils of leaves. Found in wet places, along streams, near springs into the ponderosa pine forest.

AMERICAN SPEEDWELL
Veronica americana
Latin: *verus*, truth; *icon*, image

figwort

American speedwell

II. FLOWERS STRONGLY ASYMMETRICAL (2-LIPPED); BLUE, PURPLE, WHITE, OR RED.

1. Flowers in dense spikes. Stamens 2. Plant of high altitudes.

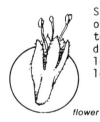
flower

Stems with woolly hairs. Leaves mostly basal, oval to oval-oblong; margins with rounded teeth. Flowers white to purple-tinged, in a dense, robust spike with several bract-like leaves below them. Sepals and petals with 4 lobes. Found in subalpine meadows.

KITTENTAILS
Besseya plantaginea
Honors Charles Bessey, American botanist
Latin: *plantaginea*, like plantain

kittentails

1. Flowers not in dense spikes. Stamens 4 or 5.

    2. Flowers club-shaped, stamens 4.

        Stems to 1 1/2 ft. Leaves narrow. Flowers club-shaped, to 2 cm long, rose-purple tinged with white. Some species parasitic on other plants. Found in the pinyon-juniper woodland.

        PURPLE-WHITE OWL-CLOVER
        Orthocarpus purpureo-albus
        Greek: *orthos*, straight; *karpos*, fruit
        Latin: *purpureus*, purple; *albus*, white

    2. Flowers not club-shaped, stamens 5.

        PENSTEMON, BEARDTONGUE
        Penstemon spp.
        Greek: *pente*, five: *stemon*, thread

        Penstemon is one of the largest genera of wildflowers. Individual species are difficult

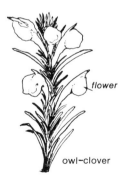
flower

owl-clover

for the beginner to identify. The genus, however, is easy to recognize. Leaves opposite, upper ones without a leafstalk. Flowers 2-lipped; the upper lip having 2, and the lower, 3 lobes. Flowers usually blue, purple, pink, red, or white. A number of species are found in the area, including:

SCARLET BUGLER *P. barbatus* var. *torreyi*. Stems to 3 ft. Leaves along the stems linear to lance-shaped, reddish, especially on the undersurface. Flowers red; lobes of the upper lip not noticeably separated. Sepals with glandular hairs, margins papery. Hummingbirds are attracted by the red flowers; these beautiful plants are called HUMMINGBIRD FLOWERS by the Indians. Found along roadsides and trails in the pinyon-juniper woodland, ponderosa pine and mixed conifer forests.

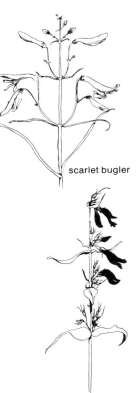
scarlet bugler

WHIPPLE'S PENSTEMON *P. whippleanus*. Stems to 3 ft, usually growing in clumps. Surfaces minutely hairy to glandular-hairy. Leaves dark green, oval to elliptical. Upper leaves oval and clasping the stem. Flowers black purple, to 3 cm long. Found in open meadows and along roadsides in mixed conifer and spruce-fir forests, subalpine meadows.

VARIEGATED PENSTEMON *P. virgatus*. Stems to 2 1/2 ft, slightly hairy. Leaves linear to lance-shaped, smooth to slightly hairy. Flowers white streaked with pink, to 2.5 cm long, all on one side of the stem. Found in pinyon-juniper woodland.

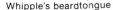
Whipple's beardtongue

BEARDTONGUE *P. secundiflorus*. Stems to 1 1/2 ft, smooth. Leaves thickened, narrowly lance-shaped. Flowers purple to pinkish purple, to 2.5 cm long. Spring bloomer. Found in dry canyons and pinyon-juniper woodland.

JAMES BEARDTONGUE *P. jamesii*. Stems to 1 1/2 ft, hairy to sticky-hairy. Leaf narrowly lance-shaped. Margins smooth to toothed. Flowers lavender, to 3 cm long, sticky-hairy on the outer surface, throat broad. Inflorescence many-flowered, all on one side of the stem. Found in dry canyons and pinyon-juniper woodland.

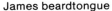
James beardtongue

beardtongue

I. FLOWERS ASYMMETRICAL, YELLOW.

1. Flowers bright yellow.

    2. Flowers two-lipped.

flowers

MONKEYFLOWER
Mimulus spp.
Latin: *mimus*, mimic; *-ulus*, small

SPOTTED MONKEYFLOWER *M. guttatus.* Stems to 2 ft hollow. Leaves oval, margins toothed. Flowers yellow, spotted with red, to 4 cm long. Indians and settlers used leaves of these species for lettuce. Found along streams and near springs.

SMOOTH MONKEYFLOWER *M. glabratus* Stems lying on ground. Leaves oval margins toothed.

spotted monkeyflower

smooth monkeyflower

  2. Flower club-shaped.

Stems to 12 cm tall. Flowers small, club-shaped. Bracts green, 3-lobed. Leaves na..ow. High mountain meadows.

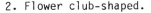

YELLOW OWL-CLOVER
Orthocarpus luteus
Greek: *orthos*, upright; *karpos*, fruit
luteus, yellow

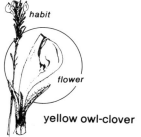

habit

flower

yellow owl-clover

1. Flowers dirty yellow.

Stems to 4 ft tall, slightly hairy. Leaves to 1 1/2 ft long, fern-like, pinnately divided nearly to the midrib vein. Flowers 3 cm long, dirty yellow, upper lip curved like a parrot's beak, lower lip with 3 lobes. Partial root parasites. Found in the mixed conifer and spruce-fir forests.

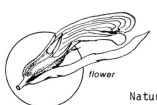

flower

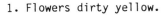

LOUSEWORT
Pedicularis grayi
Latin: *pediculus*, louse
Honors Asa Gray, professor of
Natural History, Harvard University

lousewort

IV. FLOWERS ASYMMETRICAL; COLORFUL RED, ORANGE, OR YELLOW BRACTS SURROUNDING THE FLOWER.

INDIAN PAINTBRUSH
Castilleja spp.
Honors Castillejo, Spanish botanist

Bracts surrounding the flowers more conspicuous than the flowers. Bracts either red, orange, or yellow, variously lobed. Flowers small, hidden by the bracts. Many species are parasitic on roots of other plants. Species most commonly found in the area include:

FOOTHILLS PAINTBRUSH *C. integra*. Stems to 2 ft tall, with woolly hairs. Leaves narrowly lance-shaped to linear. Bracts conspicuously orange-red, deeply lobed. Found in pinyon-juniper woodland.

SCARLET PAINTBRUSH *C. miniata*. Stems to 2 1/2 ft, with glandular hairs. Leaves linear to lance-shaped. Flowers to 2 cm long. Bracts longer than the flowers, tipped with red or vermilion. Found in ponderosa pine and mixed conifer forests.

WYOMING PAINTBRUSH *C. linariaefolia*. Stems to 2 1/3 ft, branched. Leaves narrow, thread-like. Bracts red or orange, rarely yellow. Found in canyons and mixed conifer forests.

YELLOW PAINTBRUSH *C. lineata*, Stems to 1 1/2 ft tall, woolly hairy. Leaves linear. Bracts yellowish with 3 lobes. Found in meadows of the mixed conifer forest and subalpine areas.

Indian paintbrush

flower

Nightshade Family SOLANACEAE

The flowers of different genera of the nightshade or potato family vary in shape and form, but all have five fused petals, five sepals, and five stamens. The various-colored flowers are plaited in the bud. Fruits are often berries. Several members of this family are edible, but others contain poisonous alkaloids. SACRED DATURA is an example of a highly poisonous representative. PALE WOLFBERRY *Lycium pallidum*, a shrubby species, is often found on archeological ruins (see p. 35.) Tomatoes, potatoes, and eggplants are examples of edible products of the family. Other representatives of economic importance include tobacco, peppers, petunia, and *Atropa belladona*, a source of the drug atropine.

I. FLOWERS VERY LARGE, WHITE, FUNNEL-LIKE. FRUIT SPINY.

fruit

Stems growing over 3 ft. Leaves 10 to 20 cm long, oval, uneven at the base, hairy; margins toothed to smooth. Flowers to 20 cm long, funnel-like, whitish, sometimes tinged with violet. Fruits nodding, spiny. All parts of this plant are poisonous. It is sometimes called JIMSON WEED, a corruption of the name Jamestown where soldiers stationed in 1676 were poisoned by eating the plant. It has been used ceremonially by some Indian tribes, but the poisonous alkaloids can quickly cause death. Found along trails, roadsides, and riverbanks.

SACRED DATURA
Datura meteloides
Hindi: *dhatura*, the Hindu name
Latin: like "metel" (name of unknown origin)

sacred datura

II. FLOWERS BLUE, WHEEL-SHAPED. FRUIT A BERRY.

fruit

Stems to 3 ft tall, covered with stiff hairs. Leaves to 10 cm long, oblong to lance-shaped, coarsely toothed. Flowers with 5 lobes, blue, with yellow anthers in a conical ring around the style. Berry becoming yellow-black. The Pima Indians of Arizona used the crushed berries in making cheese. It has also been used medicinally for treatment of toothache and sore throats. Found along trails and roadsides of pinyon-juniper woodland.

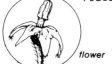
flower

SILVERLEAF NIGHTSHADE
Solanum elaeagnifolium
Latin: *solanum*, nightshade, from *solamen*, quieting (naming attributed to Pliny)
Greek: *elaia-*, resembling the olive tree
Latin: *folium*, leaf

silverleaf nightshade

III. FLOWERS WHITE OR YELLOW, WHEEL-SHAPED. FRUIT A BERRY.

Stems covered with slender, yellow prickles. Leaves oval in outline, pinnatifid. Flowers yellow, to 3 cm wide; sepals covered with prickles. Found in disturbed areas, along trails and roadsides.

BUFFALO BUR
Solanum rostratum
Latin: *rostratus*, beaked

buffalo bur

Stems to 2 ft, sparsely hairy. Leaves oval to lance-shaped, sparsely hairy. Margins smooth to toothed. Flowers white, deeply lobed, yellow spots at the base. Berry black. Along trails and roadsides.

black nightshade

BLACK NIGHTSHADE
Solanum nigrum
Latin: *niger*, black

IV. FLOWER BELL-SHAPED TO FUNNEL-LIKE. FRUIT A BERRY, SUR— ROUNDED BY AN INFLATED LANTERN-LIKE BLADDER.

Stems to 2 feet, sticky-hairy. Leaves oblong-oval to oval lance-shaped; margins toothed. Flowers to 7 mm long, yellow, spotted with blue. Sepals becoming bladder-like, resembling a small lantern enclosing the fruit. Found in disturbed areas in the pinyon-juniper woodland and ponderosa pine forest.

flower

GROUNDCHERRY
Physalis foetens var. *neomexicana*
Greek: *physalis*, bladder
Latin: *foetens*, bearing

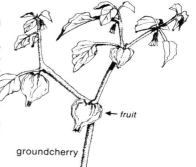
← fruit
groundcherry

Carrot Family UMBELLIFERAE

The small white or yellow flowers of plants in the parsley family are in umbrella-like clusters. Each tiny flower has five sepals, five petals, and five stamens. Stems are hollow and swollen at the base. The leafstalks (petioles) often clasp the stem. Leaves and stems in many cases are aromatic. Economically important plants in this family include food plants such as carrot, parsnip, celery, parsley; plants for flavoring such as anise, caraway, dill, chervil, fennel, lovage; ornamental plants including blue laceflower, angelica, sea holly, and cow parsnip. Some members, such as POISON WATER HEMLOCK, contain resins or alkaloids which are lethally poisonous when eaten. One should be ABSOLUTELY SURE before eating wild plants; even experts have poisoned themselves accidentally.

I. PLANTS LESS THAN 3 FT TALL, FLOWERS YELLOW, WHITE, OR PURPLE.

1. Flowers yellow.

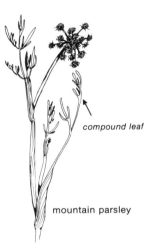
compound leaf

mountain parsley

Leaves pinnately compound, leaf divisions filament-like or lance-shaped. Leafstalk with papery or purplish margins. Flowers yellow in flat-topped clusters. Found in dry soils of the ponderosa pine and mixed conifer forests.

MOUNTAIN PARSLEY
Pseudocymopterus montanus
Greek: *psuedes*, false; *kyma*, wave
pteron, wing
Latin: *mons*, mountain

1. Flowers purple or with a club-shaped fruit.

    2. Flowers purple. Stemless.

Stemless plant, blooming very early in the spring. Leaves pinnately compound, arising from the base of plant. Flowers purple; bracts beneath flower whitish, 3-veined. Found in the pinyon-juniper woodland.

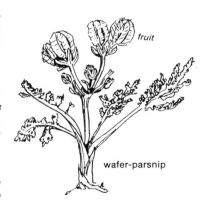
flowers
bracts
fruit
wafer-parsnip

WAFER-PARSNIP
Cymopteris bulbosus
Greek: *kyma*, wave; *pteron*, wing
Latin: *bulbus*, a bulb

    2. Plant with club-shaped fruits. Stems to 2 ft.

Delicate plant growing to 2 ft. Leaves twice pinnately compound; leaflets lance-shaped to oval. Margins coarsely toothed. Flowers blooming very early; fruits club-like, in umbrella-like clusters. Found in canyons and moist woods of the ponderosa pine and mixed conifer forests.

BLUNTSEED SWEET CICELY
Osmorhiza obtusa
Greek: *osme*, odor; *rhiza*, root
Latin: *obtusus*, blunt

II. PLANTS GROWING OVER 3 FT TALL. FLOWERS WHITE.

1. Leaves large, divided into 3 leaflets, leafstalk sheath-like.

    Stems growing to 9 ft. Leaves large, compound, divided into 3 leaflets, hairs on the undersurface. Leaf blades narrowing to a large sheath-like leafstalk which clasps the stem. Flowers in a large umbrella-like cluster 15 to 30 cm across, individual flowers small. Leaf and flower stalks of this species were used by the Indians for potherbs and basal parts of the stems as a salt substitute. It has also been used medicinally for rheumatism. Along streams in the ponderosa pine forest.

COW-PARSNIP
Heracleum lanatum
After Herakles (Hercules)
mythical hero
Latin: *lanatus*, woolly

1. Leaves 3 times divided, or once-pinnate.

    2. Stems with purple dots. Leaf veins ending at notches between teeth.

        Stems growing to 9 ft tall. One of the most POISONOUS flowering plants of the world. Can easily be confused with non-poisonous species. Leaves 1 to 3 times compound. Margins toothed. Leaf veins ending at the notches between teeth. This species was supposedly used to kill Socrates. Found along streams. Another species in the area which is highly poisonous is WATER HEMLOCK *Conium douglasii*. It looks very much like POISON HEMLOCK and is equally poisonous. It is also found along streams and ditches.

POISON HEMLOCK
Conium maculatum
Greek: *konion*, hemlock
Latin: *maculatus*, spotted

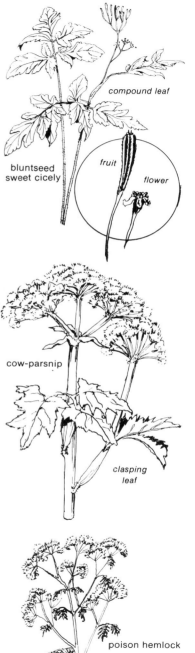

2. Stems without purple dots. Leaf veins not ending at the notches between teeth.

   3. Leaves 3 times compound

      Stems to 3 ft. Leaves 3 times compound; leaflets oval to lance-shaped. Margins pinnatifid, toothed, or smooth. Most leaves basal, few along stem. Flowers small, white, in umbrella-like clusters. Root of this species is believed to have medicinal qualities and is locally called OSHA. Moist canyons and meadows.

                    OSHA, PORTER'S LOVAGE
                      Ligusticum porteri
                    From *Liguria*, ancient name
                  for the district surrounding Genoa
                    Honors Thomas Porter (1822-1901)
                         professor of botany

   3. Leaves once-pinnate.

      Stems to 2 1/2 ft. Leaves once-pinnate; leaflets lance-shaped. Flowers white WILD PARSNIP is easily mistaken for the very POISONOUS WATER HEMLOCK *Cicuta* spp. Wild parsnips generally grow on streambanks, have fragrant rootstocks, inflorescences are hairy, and seeds are winged. Poison water hemlock generally grows in water, rootstocks are musty smelling, inflorescences are devoid of hairs; seeds are egg-shaped. Again it should be emphasized that it is easy to confuse the members of this family, many of which are poisonous. Eating wild members of the carrot family may be hazardous. Oils of a European angelica *A. archangelia* has been used in some French liqueurs. Found along streams in mixed conifer forests.

                              ANGELICA
                         Angelica pinnata
                    Latin: *angelica*, angelic
                          pinnata, feather

carrot

Porter's lovage

angelica

Valerian Family VALERIANACEAE

Flowers of the valerians are slightly irregular because they are spurred at the base. This family is closely related to the sunflower family; the sepal teeth are usually absent or in three to fifteen hair-like or feathery structures, similar to the pappi of the sunflowers. There are five lobes to the corolla and only three stamens. Leaves are opposite. Economically the family is noted for ornamentals including red valerian, corn salad, and African valerian. *Valeriana officinalis* is the source of a drug sometimes used to treat cardiac ailments.

valerian

Stems to 1 1/2 ft tall, smooth or sticky-hairy. Leaves thin with conspicuous lateral veins; the basal leaves spatula-like, the upper leaves pinnate or smooth on the margins. Flowers white to pink, in dense flat-topped clusters, slightly asymmetrical. The roots are strong-scented, giving the plant an odor resembling dirty feet. Early spring bloomer. Found in canyons of the ponderosa pine and mixed conifer forests.

flower

TOBACCO ROOT, VALERIAN
Valeriana capitata
Latin: *Valerianus*, a proper name
of unknown relation to the plant
caput, head

Vervain Family VERBENACEAE

Many members of this family have square stems, opposite or whorled leaves (leaves in a circle around the stem), and flower parts in fours or fives. Flowers are borne in spikes with many bracts. The flower looks symmetrical but often the corolla is two-lipped. The four stamens are in pairs. Teak of Southeast Asia, a valuable timber tree with hard, durable wood, is the most economically important member of this family.

VERVAIN
Verbena spp.
Latin: *verbena*, foliage

Many of the species found in the area are in disturbed soils or along roadsides. These include:

PROSTRATE VERVAIN *V. bracteata*. Stems lying on the ground. Leaves pinnately parted with 3 divisions. Flowers in a spike, light blue-purple. Found along roadsides and in disturbed ground.

VERVAIN *V. macdougalii*. Stems upright, to 3 ft tall. Leaves oblong-elliptical to oval-lance-shaped, irregularly toothed. Flowers purple, in a dense spike. Found along roadsides or in disturbed soil.

DESERT VERBENA *V. wrightii*. Stems upright, to 2 ft, hairy. Leaves bipinnatifid to tripinnatifid. Flowers rose-purple, in head-like clusters; individual blossoms salver-like, more or less 2-lipped and 5-lobed. Found in lower canyons on rocky talus slopes.

Violet Family VIOLACEAE

Violets have five petals, the lower one with a *spur* (a hollow tubular extension or saclike organ which contains nectar). This makes the flower asymmetrical. The five stamens are close to the pistil but not attached. Leaves of many species are heart-shaped. These plants are particularly abundant in moist canyons and near springs. Garden pansies and violets are domestic representatives of this family.

VIOLET
Viola spp.
Latin: *viola*, violet

Species commonly found in the area include:

I. PLANTS WITH HEART-SHAPED LEAVES.

CANADA VIOLET *V. canadensis*. Grows to 1 1/2 ft tall; stems covered with short gray or whitish hairs. Leaves heart-shaped to oval with toothed margins. Flowers white, sometimes with purplish veins, petals 6 mm to 1.5 cm long. Found in moist canyon bottoms of the ponderosa pine and mixed conifer forests.

Canada violet

NORTHERN BOG ORCHID *V. nephrophylla*. A stemless violet growing to 24 cm tall. Flowering stalk smooth. Leaves heart-shaped to kidney-shaped. Margins with rounded teeth. Flowers purple to violet, petals to 1.5 cm with hairs; spurred. Moist canyons and near springs.

WESTERN DOG VIOLET *V. adunca*. Stems growing to 25 cm, smooth to hairy. Leaves round to oval. Margins with rounded teeth. Flowers purple or violet, petals to 1.2 cm; lateral petals hairy. Found in moist canyons and on mountain slopes.

II. PLANTS WITH PALMATELY DIVIDED LEAVES

LARKSPUR VIOLET *V. pedatifida*. Stemless violet. Flowering stalks to 20 cm tall. Leaves palmately divided. Midrib vein hairy. Flowers purple to 2 cm long; lateral petals hairy. Found in moist meadows.

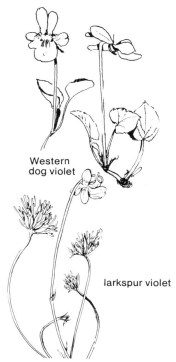
Western dog violet

larkspur violet

Caltrop Family ZYGOPHYLLACEAE

Caltrops are either herbaceous perennials or shrubs. Leaves are compound and opposite. Flowers have five sepals, five petals, and many stamens. Members of the family include LIGNUM VITAE *Guaiacum officinale*, having the hardest and most dense of any wood, and CREOSOTE BUSH *Larrea divaricata*, which is abundant in areas south of Albuquerque. Our local representative is a noxious weed.

Stems trailing. Leaves pinnately compound; leaflets 5 to 7, oblong to elliptical. Flowers yellow, to 1 cm. A noxious weed whose spiny fruit is a menace to bicycle tires. Naturalized from Europe. Generally found in disturbed soil, particularly sandy soil.

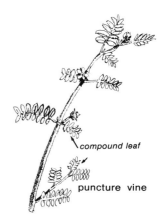

BURNUT, GOATHEAD, PUNCTURE VINE
Tribulus terrestris
Greek: *treis*, three; *belos*, dart
Latin: *terra*, earth

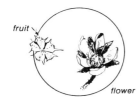

habit

GLOSSARY

alpine: found on mountains above timberline.

alternate: leaves occurring singly, not in pairs or whorls. Compare *opposite*.

ament: same as *catkin*.

annual: living and growing for only one season; completing the life cycle in one season and then dying.

anther: the pollen-bearing part of a stamen.

areole: the raised area on a cactus, bearing small spines.

asymmetrical flower: flower having only two identical halves (mirror images) about one plane. Compare with *symmetrical*.

awn: a slender bristle, usually at the tip of a stem, leaf, or bract.

axil: the upper angle between the leafstalk or flower stalk and the stem.

banner: the upper, larger petal in a flower of the pea family. See *papilionaceous*.

barbed: having a backward-projecting tip; like a fishhook.

basal leaf: leaf at the base of a plant.

biennial: living two years; completing the life cycle in two seasons. The first season vegetative; the second season devoted to seed production.

bipinnate leaves: leaves two times pinnate. Compare *pinnate*.

blade: the portion of a leaf other than the leafstalk; generally flattened to provide for collection of sunlight and gases for photosynthesis, the process of food production.

bract: a modified leaf usually below a flower or group of flowers; in grasses, a modified flower part; in conifers a woody or papery part of the cone.

bristle: stiff, hair-like structure.

bud: the developing flower or leaf, usually protected by scales.

bulb: a short, swollen, underground stem used for food storage.

bur: a rough, spiny, or prickly seed or fruit, usually hooked or barbed.

calyx: the collective term for the sepals, the outer series of the flower structure; leaf-like, usually green, but in some species colored. Compare with *corolla*.

capsule: a dry fruit made of several compartments containing one or more seeds. May or may not split open when mature.

catkin: also called ament. Elongate cluster of flowers without petals, wind pollinated; usually of early blooming trees. (Catkin literally means a kitten, apparently first used in 1578 to describe a pussywillow.)

chaff: the scale-like structures on the receptacle of many composites. Each disk and ray flower is surrounded by these structures, which remain after the seeds have fallen. Also the dry flower parts of grasses.

chlorophyll: pigment giving plants green color; important pigment for photosynthesis.

clasping leaf: a leaf with leafstalk partially wrapped around the stem.

claw: the narrowed base of a petal.

composite: a member of the Compositae, or sunflower family.

compound leaf: a leaf completely separated into two or more leaflets; leaflets may be arranged pinnately or palmately. Compare with *simple*.

cone: dry fruit having overlapping, woody scales or bracts covering the maturing seeds.

conifer: evergreen, cone-bearing tree such as pine, spruce, fir.

corolla: the collective term for the petals, the second series in flower structure; usually colored; occasionally missing. Compare *calyx*.

corona: See *crown*.

cotyledon: a seed leaf; a starchy food storage structure surrounding the developing plant embryo.

crown: a petal-like structure on milkweed flowers appearing like a crown, technically a corona.

deciduous: shedding leaves at a specific season or stage in the growth of a plant; not persistent or evergreen. Compare *evergreen*.

dicot: a shortened name for a dicotyledonous plant; a plant having two cotyledons or seed leaves; example: beans. See *cotyledon*. Compare *monocots*.

disk: the central part of the head of a composite, consisting of symmetrical, tubular flowers. Some flowers such as thistles have only disk flowers.

dissected leaf: leaf cut or divided into many narrow segments.

elliptical leaf: leaf shaped like an ellipse, wide in the middle with the two narrow ends equal.

entire: leaf or petal margins smooth, without teeth.

evergreen: bearing green leaves throughout the year.

exerted: protruding beyond.

female flowers: flowers having only pistils or seed-producing organs.

fern-like leaf: leaf which is dissected or divided into narrow segments.

fetid: having a disagreeable odor.

floret: small flower, often part of a dense cluster; usually referring to grasses.

free-petal flower: flowers having petals that are each a separate unit, not fused in any way, and falling off separately.

frond: the leaf of a fern.

fruit: ripened ovary of seed-bearing plants, containing seeds.

glandular: bearing glands, that is, swellings which contain some secretions, usually on hairs.

head: a dense, compact cluster of flowers, as in sunflowers or clover.

herbaceous: having no persistent woody stem above the ground.

hip: fleshy, berry-like, often colorful fruit of a rose.

hooked: abruptly curved at the tip. Compare with *barbed*.

hybridization: formation of a crossbreed, offspring of genetically dissimilar parents.

hypanthium: an elongation of the flower beneath the sepals.

internode: portion of the stem between leafstalks.

introduced: pertaining to plants brought into the country from another country either for purposes of cultivation or unintentionally as a hitchhiker in cargo.

involucre: a circle of leaf-like bracts or scales under a flower head or inflorescence, particularly in composites and clovers.

keel: the fused lower petals of a flower of the pea family, resembling the prow of a boat. See *papilionaceous*.

kidney-shaped leaf: a leaf longer than wide, widest part closes to the base of the leaf; gently tapering toward both ends.

lateral: on the side.

leafstalk: the stalk by which the leaf blade is attached to the stem. See *compound leaf*.

legume: characteristic fruit of the pea family; pod splitting into two halves with seeds attached to the lower side of one half.

lenticel: small pore on the surface of stems of woody plants, allowing gases to enter and leave the tissue.

linear leaf: narrow, flat leaf with parallel sides, very elongated, as in grasses.

lip: either the upper or lower portion of an asymmetrical flower.

lobed: having lobes, rounded projections, on leaf blades or on petals.

lyre-like leaf: pinnatifid leaf with terminal segments large and rounded and lower lobes small. See *pinnate*, *pinnatifid*.

male flowers: flowers having only stamens, the pollen-bearing organs.

many: more than ten of any structure.

membranous: thin, papery.

midrib: the main or central rib of a leaf.

monocots: shortened name for monocotyledonous plant; a plant having one seed leaf; example: corn. See *cotyledon*. Compare *dicot*.

needle-like: long, slender, rather rigid.

nerve: a simple, unbranched vein of a leaf.

net-veined leaf: leaf having veins in a network, rather than parallel.

node: the place on the stem where leaves or branches are attached.

oblong leaf: leaf having sides nearly parallel, length less than ten times longer than width.

odd-pinnate: a pinnately compound leaf with a single terminal leaflet. Having an odd number of leaflets. See *pinnate*.

opposite leaves: leaves attached two at a node, situated across the stem from each other. Compare *alternate*.

oval: used loosely for broadly elliptical or egg-shaped.

overstory: the tallest plants of an area, usually referring to trees or shrubs.

palmate leaf: leaf having lobes or veins originating from one place, like the fingers growing from the palm of the hand. Compare *pinnate leaf*.

papery: thin, usually whitish, like paper.

papilionaceous flower: a pea-type flower with a banner, wings, and a keel.

pappus: a tuft of bristles or bracts surrounding the seed of certain members of the sunflower family.

parallel-veined leaf: leaf having veins running parallel to each other. Compare *net-veined*.

parasite: an organism growing upon and obtaining nourishment from another organism.

perennial: a plant living for three or more years; roots not dying over the winter.

perianth: the flower parts consisting of petals and sepals.

petal: one of the parts of a flower; situated inside the sepals, variously colored.

petiole: the leafstalk attaching the leaf to the stem; may or may not be present. If not present, the leaf is sessile.

phyllary: the special name for bracts below a head type of inflorescence, such as in the sunflower family, collectively called an involucre.

pinna: one of the divisions of a fern leaf.

pinnate leaf: a compound leaf with leaflets on two opposite sides of an elongated axis, feather-like. Compare *palmate leaf*.

pinnatifid leaf: leaf that is pinnately lobed, cleft or parted half way to the midrib vein.

pistil: seed-producing organ of a flower; develops into the fruit.

pith: spongy center of a stem.

pollen: the male spores in the anther.

prickle: a small, slender outgrowth of young bark, coming off with the bark. Compare with *spine, thorn*.

prostrate: lying on the ground.

ray flowers: the strap-shaped flowers of the sunflower family.

receptacle: the expanded portion of the flower stalk that bears the organs of the flower or collections of flowers.

riparian: pertaining to banks of rivers, streams, and ponds.

rosette: a dense cluster of leaves arranged in a circular fashion at the base of the plant.

salver-like flower: corolla with a long, slender tube and a flattened top.

samara: a winged fruit which does not split open; usually of elms and maples.

saprophyte: a plant which lives on dead and decaying material, usually lacking green color.

scorpion-like: coiled like the tail of a scorpion.

seed: a mature ovule, consisting of the embryo and food.

sepal: one of the parts of the flower, usually green in color. Outermost series in floral structure.

silky: having long, soft hairs.

simple leaf: leaf having only one part, not divided into leaflets. Compare *compound leaf*.

smooth: when referring to surfaces, without hairs; when referring to margins, without teeth.

sorus: (pl. sori.): a cluster of spore-containing sacs on the fern frond.

spathe: a large bract or bracts sheathing or enclosing an inflorescence, as in the spiderwort family.

spatula-like leaf: leaf that is broad at the tip and narrow at the base.

spike: an elongated cluster of flowers attached directly to the stem, like a cattail.

spine: a sharp-pointed, rigid, deep-seated outgrowth from the stem, not pulling off with the bark.

spore: a reproductive cell of a fern that develops into a plant without union with other cells.

stamen: the male organ of a flower, bearing pollen in the anther.

staminode: a stamen without an anther.

star-shaped: branches of hairs radiating from a common center.

stigma: part of the pistil receiving pollen, can be one to several.

stipe: the stem of a fern frond.

style: the portion of the pistil connecting the ovary with the stigma.

subshrub: a small, low-growing shrub.

symmetrical flower: flowers having two identical halves (mirror images) when bisected through any plane. See *asymmetrical*.

tendril: organ of support; coiling outgrowth of the stem or leaf.

thorn: a stiff, hard, sharp-pointed structure. Usually a modified leaf or branch.

twining: climbing by coiling around a support.

two-lipped flowers: asymmetrical flowers having petals arranged in two major extensions, one above the other.

umbrella-like flower head or cluster: a flat-topped inflorescence, all branches originating at one point.

understory: the undergrowth or shorter plants of an area.

united petals: petals not separate but all fused into a unit and all coming off as a unit.

wing: the side extensions on a flower of the pea family. See *papilionaceous*.

woolly: having long, soft, interwoven hairs.

SELECTED REFERENCES

Agricultural Research Service, US Department of Agriculture, 1971 *Common Weeds of the United States*. Dover Publications. NY.

Benson, L. 1969. *The Cacti of Arizona*. Univ. of Ariz. Press. Tucson. Arizona.

Craighead, J. F., F. C. Craighead, and R. J. Davis. 1963. *A Field Guide to Rocky Mountain Wildflowers*. Houghton Miffin Co. Boston. Mass.

Curtin, L. S. M. 1965. *Healing Herbs of the Upper Rio Grande*. Southwest Museum. Los Angeles, Calif.

U. S. Forest Service. 1937. *Plant Range Handbook*. US Goverment Printing Office. Washington, D. C.

Foxx, T. S. and G. D. Tierney. 1980. "Status of the Flora of the Los Alamos National Environmental Research Park. LA-8050-NERP, Vol. 1 Los Alamos National Laboratory Report.

Foxx, T. S. and G. D. Tierney. 1984. "Status of the Flora of the Los Alamos National Environmental Research Park, A Historical Perspective. LA-8050-NERP, Vol. 2. Los Alamos National Laboratory Report.

Foxx, T. S. and G. D. Tierney. 1984. Checklist of Vascular Plants of the Pajarito Plateau and Jemez Mountains. LA-8050-NERP. Vol. 3 Los Alamos National Laboratory Report.

Gould, F. W. 1977. *Grasses of the Southwestern United States*. Univ. of Ariz. Press, Tucson, Arizona.

Harrington, H. D. 1974. *Manual of Plants of Colorado*. Sage Books. Chicago, Ill.

Harrington, H. D. and Y. Matsumura. 1974. *Edible Native Plants of the Rocky Mountains*. Univ. of New Mex. Press. Albuquerque, New Mex.

Kirk, D. R. 1970. *Wild Edible Plants of the Western States*. Naturgraph Publishers. Healdsburg, Calif.

Martin, W. C., C. R. Hutchins and R. G. Woodmansee. 1970. *A Flora of the Sandia Mountains*. Univ. of New. Mex. Biology Department. Albuquerque, New Mex.

Martin, W. C. and C. R. Hutchins. 1981. *A Flora of New Mexico*. Two Volumes. J. Cramer, Germany.

Moore, M. 1979. *Medicinal Plants of the Mountain West*. Museum of New Mexico Press. Santa Fe, New. Mex.

Nelson, R. 1969. *Handbook of Rocky Mountain Plants*. Dale Stuart King. Tucson, Arizona.

PRONUNCIATION GUIDE

Scientific names are derived in two ways: 1) Latin or Greek roots are combined to form a word indicating something about the organism and 2) the latinizing of names of people, places, or things. In the body of this manuscript, the derivation of each scientific name has been given. Here is presented some basic information about their pronunciation. Even though the names are from Latin or Latinized the way they are spoken does not always follow the rules of pure Latin but have been anglicized. There is often disagreement between botanists as to proper pronunciation. However, a knowledge of rules of Latin can be a guide. Presented below are some of these rules, a pronunciation key, a listing of some scientific names used in this book, and their pronunciation as found in *The New Pronouncing Dictionary of Plant Names*, *Latin for Taxonomists*, and *Shrubs and Trees of the Southwest Uplands*.

Rules

1) Accents: If a word has only two syllables, the accent falls on the first syllable. When a word has more than two syllables, the accent may fall on the next-to-last or third-from-last syllable.

2) Vowels: Vowels may be pronounced with either long or short sounds.

3) Diphthongs: A number of vowels are written together and pronounced as one sound. See the Pronunciation Key for pronunciation of the various diphthongs.

4) Consonants: The various consonants have different sounds.

    c is pronounced as k in cat before a, o, oi, or u
    c is pronounced as s in sent before ae, i, e, oe, or y
    ph is pronounced as f
    ch is pronounced as k in chorus
    x is pronounced as z when it is the initial letter
    The first letter of paired consonants is silent, i. e., *Pseudotsuga*.

5) Family names usually end in -aceae, pronounced as a-see-ee. In this book some family names end in -ae, pronounced -ee. In some books, the families whose names end in -ae have been given names ending in -aceae to make them more consistent. Below are listed the family names ending in -ae in this book and the corresponding names ending in -aceae.

                                Grass family Gramineae-Poaceae
                                Mint family Labiatae-Lamiaceae
                           Mustard family Cruciferae-Brassicaceae
                         Sunflower family Compositae-Asteraceae

6) Synonomy: Synonyms are discarded names for the same plant. In this book *A Flora of New Mexico* by Martin and Hutchins was used as the authority for scientific names. Other books or recent monographs may assign different names to the plants listed in this publication. Since the goal of the author was to have this book be a beginning text, it was determined that consistency of terminology with existing flora was important for the beginning student. For information on synonomy see *A Checklist of Vascular Plants of the Pajarito Plateau and Jemez Mountains* by Foxx and Tierney.

Pronunciation Key

When the vowel or diphthong sounds like the word indicated it will be written in the manner to the right of the column. Capitalized letters within the pronunciation guide indicate the accent.

```
a as in fat..............a              DIPHTHONGS
a as in fate.............ay
a as in far..............ah        ae as in Caesar.........ee
a as in fall.............ah        ai as in rail...........ay
a as in Persia...........uh        au as in auk............aw
a as in fare.............eh        ei as in height.........eye
e as in met..............eh        eu as in feud...........yew
e as in me...............ee        ia as in Asia...........uh
e as in her..............ur        oe as in bee............ee
i as in pin..............ih        ou as in soup...........oo
i as in pine.............eye       ui as in ruin...........ew
o as in not..............ah        oi as in oil............oy
o as in note.............oh
o as in move.............oo
o as in nor..............o
u as in tub..............uh
u as in mute.............yew
y as in symbol...........ih
y as in by...............eye
```

Abies (AY-beez)
acanthus (ay-KAN-thus)
acaulescent (ay-kaw-LEHS-ehnt)
acaulis (ay-KAWL-ihs)
Acer (AY-sur)
Achillea (ak-ih-LEE-ah)
Aconitum (ak-oh-NEYE-tuhm)
Actaea (ak-TEE-ah)
Adianthum (ay-dee-AN-thum)
Agrostis (ah-GRAHS-tihs)
Ailanthus (ay-LAN-thus)
alba (AL-bah)
albescens (al-BEHS-enz)
albicans (AL-bihk-anz)
Allium (AL-ee-uhm)
alnifolia (uhl-nih-FOHL-lih-uh)
Alnus (AL-nus; UHL-nuhs)
alpestris (al-PEHS-trihs)
altissima (al-TIHS-ih-mah)
Amaranthus (am-ah-RAN-thuhs)
Amelanchier (am-eh-LAN-kih-ehr)
americana (ah-mehr-ih-KAY-nah)
Amorpha (ah-MOR-fah)
andromeda (an-DRAHM-eh-dah)
Andropogon (an-droh-POH-gahn)
Androsace (an-DROH-saye-seh)
Amenome (ah-NEHM-oh-nee)
Angelica (an-JEHL-ih-kah)

angustifolium (an-guhs-tih-FOHL-lih-uhm)
annuus (AN-yew-ahs)
Antennaria (an-teh-NEH-rih-uh)
antirrhinum (an-tih-REYE-nuhm)
aquaticum (ah-KWAT-ih-kuhm)
aquifolium (ak-wih-FOHL-lih-uhm)
Aquilegia (ak-wih-LEE-jih-uh)
Arabis (AR-ah-bihs)
Aralia (ah-RAY-lih-uh)
arborescens (ahr-bo-REHS-ehnz)
Arctostaphylos (ahr-toh-STAF-ih-lohs)
Arenaria (ar-eh-NAY-rih-ah,
    ar-eh-NEH-rih-ah)
argentea (ahr-jehn-TEE-ah)
Artemisia (ahr-TEH-meez-ih-uh,
    ahr-TEE-mihsh-ih-uh)
arvensis (ahr-VEHN-sihs)
Asclepias (as-KLEE-pih-as
    as-KLEH-pih-as)
asperifolia (as-pehr-ih-FOHL-ih-uh)
asperula (as-PUR-yew-lah)
Aster (AS-tur)
Astragalus (as-TRAG-ah-luhs)
Atriplex (AT-rih-plehks)
atropurpurea (at-ROH-pur-pur-ee-ah)
aurantiacus (aw-ran-tih-AY-kuhs)
avicularis (ah-vihk-yew-LA-rihs)
barbatus (bahr-BAY-tuhs)

Berberis (BUR-bur-ihs)
Betula (BEHT-yew-lah;
    BEHCH-yew-lah)
Bidens (BEYE-dehnz)
bifida (BEYE-fih-dah)
bipinnata (beye-pin-AH-tah)
Bromus (BROH-muhs)
Cactaceae (kak-TAY-see-ee)
Calochortus (kal-oh-KOR-tuhs)
Calypso (kah-LIHP-soh)
Campanula (kam-PAN-yew-lah)
Campanulaceae (kam-pan-yew-LAY-see-ee)
Castilleja (kahs-tih-LEH-yah)
Ceanothus (see-ah-NOH-thuhs)
Celtis (SEHL-tihs)
Cerastium (sur-RAS-tih-uhm)
Cercocarus (sur-koh-KAHR-puhs)
Chenopodium (kee-noh-POH-dih-uhm)
Chimaphila (keye-MAHF-ih-lah)
Chrysopsis (krihs-AHP-sihs)
Cichorium (sih-KOR-rih-uhm)
clavata (klah-VAH-tah)
Clematis (KLEHM-a-tihs)
Cleome (klee-OH-mee)
coccinea (kok-SIH-nee-ah)
Commelina (kohm-eh-LEYE-nah)
compactus (kahm-PAK-tuhs)
Compositae (kahm-PAHZ-ih-tee)
conoides (ko-NOY-deez)
Convolvulus (kon-VOHL-vuh-luhs)
Corydalis (koh-RIHD-ah-lihs)
Corypantha (kohr-ih-FAN-thah)
Cosmos (KAHZ-mohs)
Crataegus (krah-TEE-guhs)
Crepis (KREE-pihs)
Croton (KROH-tahn)
Cruciferae (kroo-SIF-ur-ee)
Cucurbita (kew-KUR-bih-tah)
Cucurbitaceae (kew-kur-bih-TAY-see-ee)
Dactylis (DAK-tih-lihs)
Datura (dah-TUR-rah)
decumbens (dee-KUHM-behnz)
Delphinium (dehl-FIHN-ih-uhm)
divergens (deye-VUR-jehnz)
Dodecatheon (doh-de-KATH-ee-ahn)
Draba (DRAHB-bah)
Dracocephalum (dray-koh-SEHF-ah-luhm)
Echinocactus (ee-keye-noh-KAK-tuhs)
Echinocereus (ee-keye-noh-SEE-ree-uhs)
Elaeagunus (ee-lee-AG-nuhs)
elegans (EHL-eh-ganz)
Epilobium (ehp-ih-LOH-bih-uhm)
Epipactis (ehp-ih-PAK-tihs)
Equisetum (ehk-wih-SEE-tuhm)

Eragrostis (ehr-ah-GRAHS-tihs)
Ericaceae (ehr-ih-KAY-see-ee)
Eriogonum (eh-ri-AHG-oh-nuhm)
Erodium (eh-ROH-dee-uhm)
Erysimum (eh-RIHS-ih-muhm)
Eupatorium (yew-pah-TORIH-uhm)
Euphorbia (yew-FOR-bih-uh)
filix-femina (fee-lihks-FEHM-ih-nah)
floribunda (flor-ih-BUHN-dah)
Fragaria (frah-GEH-rih-uh)
fragilis (FRAJ-ih-lihs)
fruticosa (froo-tih-KOH-sah)
Gaillardia (gay-LAHR-dih-uh)
Galium (GAY-lih-uhm)
Gentiana (jen-shih-AH-nah)
Geranium (jeh-RAY-nih-uhm)
Geum (JEE-uhm)
Gilia (GIHL-ih-uh)
glauca (GLAW-kah)
Gramineae (grah-MIHN-ee)
Grindelia (grihn-DEL-lih-uh,
    grih-DEE-lih-uh)
guttatus (gyew-TAY-tuhs)
Habenaria (hab-en-NAY-rih-ah)
Hedeoma (hee-dee-OH-mah)
Helenium (heh-LEE-nih-uhm)
Helianthus (hee-lih-AHN-thuhs)
Heracleum (hur-ah-KLEE-uhm)
Heuchera (YEW-kehr-ah)
Holodiscus (hol-oh-DIHS-kuhs)
Hordeum (HOR-dee-uhm)
Humulus (HYEW-myew-lihs)
Hypericum (heye-PEHR-ih-kuhm)
hystrix (HIHS-trihks)
Ipomoea (ihp-oh-MEE-ah;
    eye-poh-MEE-ah)
Iris (EYE-rihs)
Jamesia (JAYMZ-ih-uh)
Juncus (JUHN-kuhs)
Juniperus (joo-NIHP-ur-uhs)
Kochia (KOH-kih-uh; KOH-chih-uh)
Labiatae (lay-bih-AY-tee)
Lactuca (lak-TOO-kah)
Lathyrus (LAHTH-ih-ruhs)
latisquamus (lat-ih-SKWAY-muhs)
Leguminosae (leh-gyew-mih-NOH-see)
Lepidium (leh-PIDH-ih-uhm)
Liatris (leye-AY-trihs)
Lilium (LIHL-ih-uhm)
Linum (LEYE-huhm)
Lithospermum (lihth-oh-SPUR-muhm)
Lobelia (loh-BEE-lih-uh)
Lolium (LOH-lih-uhm)
Lonicera (lo-NIHS-ur-ah)

Lotus (LOH-tuhs)
Lupinus (lew-PEYE-nuhs)
luteus (LEW-tee-uhs)
Lycium (LIHS-ih-uhm)
Macracanthus (mak-rah-KAN-thuhs)
Malva (MAL-vah)
Marrubium (ma-REW-bih-uhm)
Medicago (mehd-ih-KAY-goh,
    med-ih-KAH-goh)
Melilotus (mehl-ih-LOH-tuhs)
Mentha (MEHN-thah)
Mentzelia (mehnt-ZEE-lih-uh)
Mertensia (mur-TEHN-sih-uh)
micranthus (meye-KRAHN-thus)
Mimulus (mihm-YEW-luhs)
Mirabilis (mih-RAB-ih-lihs)
Monarda (moh-NAHR-dah)
Monotropa (mah-NOHT-roh-pah)
novae-angeliae (noh-vay-an-GEH-lih-ee,
    noh-vee-an-GEH-lih-ee)
oblongifolia (ahb-long-ih-FOH-lih-uh)
occidentalis (ak-sih-dehn-TAY-lihs)
Oleaceae (oh-lee-AY-see-see)
Opuntia (oh-PUHN-shi-ah)
Orchidaceae (or-kih-DAY-see-ee)
Orthocarpus (or-thoh-KAHR-puhs)
Pachystima (pah-KIHS-tih-mah)
papilionaceous (pah-pihl-ih-oh-NAY-shuhs)
Parthenocissus (pahr-then-oh-SIHS-uhs)
Pedicularis (peh-dihk-yew-LAY-rihs)
Pellaea (peh-LEE-ah)
Penstemon (pen-STEE-mohn,
    pen-STEH-mohn)
Petalostemum (peht-ah-loh-STEE-muhm)
petiolaris (peht-ih-oh-LAY-rihs)
Phacelia (fah-SEE-lih-uh)
philadelphus (fihl-ah-DEHL-fuhs)
Phleum (FLEE-uhm)
Physalis (FIH-say-lihs, FIS-ah-lihs)
Physocarpus (feye-soh-KAHR-puhs)
Plantago (plan-TAY-goh)
Polemonium (pahl-eh-MOH-nih-uhm)
Polygonum (pahl-LIH-goh-nuhm)
Polypodium (pahl-ih-POH-dih-uhm)
Portulaca (pohr-tew-LAH-kah)
Potentilla (POH-ten-tihl-ah)
Prunella (proo-NEHL-ah)
Prunus (PROO-nuhs)
Pseudotsuga (soo-doh-SOO-gah)
Ptelea (TAY-lee-ah)
Pteridium (teh-RIHD-ih-uhm)
pubescens (pew-BEHS-ehnz)
pulchellus (puhl-CHEHL-uhs)
pumilus (PEW-mih-luhs)

Pyrola (PEYE-roh-lah, PIR-oh-lah)
Quercus (KWUR-kuhs)
quinquefolia (kwihn-kweh-FOH-lih-uh)
Ranunculus (rah-NUHNG-kew-luhs)
repens (REE-pehnz)
reticulata (reh-tihk-yew-LAY-tah,
    reh-tihk-yew-LAH-tah)
Rhus (ROO-z)
Ribes (REYE-beez)
Robinia (roh-BIHN-ih-uh)
Rosa (ROH-zah)
Rosaceae (roh-ZAY-see-ee)
rotundifolia (roh-tuhn-dih-FOH-lih-uh)
Rubus (ROO-buhs)
Rudbeckia (rood-BEHK-ih-uh)
Rumex (ROO-mehks)
Salix (SAY-lihks)
Salvia (SAHL-vih-uh; SAL-vih-uh)
Sambucus (sam-BEW-kuhs)
Saxifraga (sahks-IHF-ray-gah)
scoparia (skoh-PAH-rih-uh)
Senecio (seh-NEE-see-oh)
Sidalacea (seye-DAHL-see-ah)
Sisyrinchium (sihs-ih-RIHNG-kih-uhm)
Smilacina (smeye-lah-SEE-nah)
Solanum (soh-LAH-nuhm)
Solidago (soh-lih-DAH-goh)
Spheralcea (sfee-RAL-see-ah)
Stellaria (steh-LAY-rih-uh)
stipa (STEYE-pah)
Symphoricarpos (sihm-foh-rih-KAHR-pohs)
tagetes (tah-JEE-teez)
Tamarix (TAM-ah-rihks)
Taraxacum (tah-RAK-sah-kuhm)
tectorum (tehk-TOH-rhum)
tenuifolius (ten-yew-ih-FOH-lih-uhs)
Thalictrum (thah-LIHK-truhm)
Thelesperma (thel-eh-SPUR-mah)
Thermopsis (thur-MOP-sihs)
Thlaspi (THLAS-pih, THLAS-pee)
Tragopogon (trag-oh-POH-gahn)
trilobata (treye-loh-BAY-tah)
Typha (TEYE-fah)
Urtica (UR-tih-kah)
uva-ursi (uh-vah-UR-seye)
Vaccinium (vak-SIHN-ih-uhm)
Valeriana (vah-lee-rih-AH-nah)
Veratrum (veh-RAY-truhm)
Verbena (vur-BEE-nah)
Veronica (veh-ROHN-ih-kah)

INDEX

Abies
    concolor, 19
    lasiocarpa, 19
Acer
    glabrum, 23
    negundo, 22
ACERACEAE, 22-23
Achillea lanulosa, 93
Aconitum columbianum, 165
Actaea arguta, 34
actinea, 92
adder's mouth, 105
Agoseris
    aurantiaca, 75
    glauca, 75
Agrimonia striata, 170
agrimony, 170
Agropyron
    desertorum, 54
    smithii, 57
    trachycaulum, 57
Ailanthus altissima, 22
alder, thinleaf, 26
alfalfa, 147
alfileria, 139
alligator juniper, 20-21
Allium
    cernuum, 103
    geyeri, 103
Alnus tenuifolia, 26
alumroot, 173
AMARANTHACEAE, 108, 110
Amaranth family, 108, 110
Amaranthus
    graecizans, 108
    retroflexus, 110
Ambrosia coronopifolia, 81
Amelanchier utahensis, 43
American
    hop, 12
    speedwell, 175
Amorpha fruticosa, 35
ANACARDIACEAE, 32, 33
Anaphalis margaritacea, 84
Andropogon
    gerardii, 57
    scoparius, 64

Androsace septentrionalis, 164
anemone, candle, 168
Anemone cylindrica, 168
angelica, 183
Angelica pinnata, 183
anoda, 154
Anoda cristata, 154
antelope horns, 122
antelope sage, 163
Antennaria parvifolia, 84
Apache plume, 41
APOCYNACEAE, 121
Apocynum
    androsaemifolium, 121
    cannabinum, 121
Aquilegia
    caerulea, 166
    elegantula, 166
    triternata, 166
Arabis fendleri, 133
Arceuthobium vaginatum, 49
Arctostaphylos uva-ursi, 28-29
Arenaria fendleri, 129
Aristida
    arizonica, 60
    divaricata, 60
    longiseta, 60
Arizona peavine, 149
Artemisia
    bigelovii, 79-80
    carruthii, 79-80
    dracunculus, 79-80
    filifolia, 47
    franserioides, 79-80
    frigida, 79-80
    ludoviciana, 79-80
    tridentata, 42, 80
ASCLEPIADACEAE, 122-123
Asclepias
    asperula, 122
    latifolia, 122
    speciosa, 122
    subverticillata, 123
    tuberosa, 123
aspen, 24
aster
    Bigelow, 96
    marsh, 97
    New England, 97
    sand, 95
    smooth, 97

    Townsend's, 95
Aster
    ericoides, 96
    hesperius, 97
    laevis, 97
    novae-angliae, 97
Astragalus
    missouriensis, 150
    praelongus, 150
Athyrium filix-femina, 73
Atriplex canescens, 46

Bahia dissecta, 91
baneberry, 34
barberry
    creeping, 28
    Fendler, 36
Barberry family, 28, 36
barnyard grass, 59
bearberry, 28-29
beardtongue, 176
bedstraw
    fragrant, 172
    northern, 172
Beech family, 23-24
beeplant, Rocky Mountain, 127
beggarticks, 94
bellflower, 126
Bellflower family, 126
BERBERIDACEAE, 28, 36
Berberis
    fendleri, 36
    repens, 28
berlandiera, 91
Berlandiera lyrata, 91
Besseya plantaginea, 175
BETULACEAE, 26-27
Betula occidentalis, 27
Bidens bigelovii, 94
bindweed
    black, 15
    field, 15-16, 130
Birch family, 26-27
birch, water, 26-27
bittercress, 131
bitterweed, 92
black bindweed, 15
black-eyed Susan. 88
blackfoot, 95
bladderpod, 134
Blazing star family, 152

blazing star, white-stemmed, 152
Blepharoneuron tricholepis, 63
bluebells
    franciscan, 124
    Parry, 126
blue-eyed grass, 100
blue flag, 99
bluestem
    big, 57
    little, 64
bluntseed sweet cicely, 181-182
bog orchid, 105
Borage family, 124-125
BORAGINACEAE, 124-125
Bouteloua
    curtipendula, 56
    eriopoda, 58
    gracilis, 58
    hirsuta, 58
boxelder maple, 22
bracken fern, 72
bricklebush, California, 78
Brickellia
    californica, 78
    grandiflora, 78
bristlegrass green, 54-55
brittle fern, 72
bromegrass
    mountain, 61
    nodding, 61
    smooth, 61
Bromus
    anomalus, 61
    ciliatus, 61
    inermis, 61
    marginatus, 61
    tectorum, 62
Broomrape family, 50
Buchloe dactyloides, 53
buckbrush, 35, 37
Buckwheat family, 15, 109, 162-163
buckwheat, 162-163
Buffalo bur, 180
Buffalo
    grass, 53
    false, 52
bursage, 81
burnut, 187
Buttercup family, 12-13, 34, 165-168
buttercup
    heart-leaved, 166-167
    homely, 166-167
    Macoun's, 166-167
butterflyweed, 123
butterweed
    New Mexico, 90
    notchleaf, 89

CACTACEAE, 68-71
cactus
    pincushion, 71
    starvation, 69
Cactus family, 68-71
Calochortus
    gunnisonii, 101
    nuttallii, 101
Calylophus hartwegii, 157
Calypso bulbosa, 105
Caltrop family, 187
Campanula
    parryi, 126
    rotundifolia, 126
CAMPANULACEAE, 126
Canada wildrye, 55
cancer-root, 50
candle anemone, 168
candytuft, wild, 132
canyon grape, 13
Caper family, 127
CAPPARIDACEAE, 127
CAPRIFOLIACEAE, 33, 39, 40
Capsella bursa-pastoris, 133
Cardamine cordifolia, 131
Carex
    festivella, 67
    praegracilis, 67
Carrot family, 181-183
CARYOPHYLLACEAE, 128
Castilleja
    integra, 178
    linariaefolia, 178
    lineata, 178
    miniata, 178
catchfly, 128
catsfoot, 84
cattail, 65
Cattail family, 65
Ceanothus fendleri, 35, 37
CELASTRACEAE, 29
Celtis reticulata, 27
Cenchrus pauciflorus, 53
Cerastium arvense, 128
Cercocarpus montanus, 38, 42-43
chamisa, 46-47
cheeseweed, 154
CHENOPODIACEAE, 45, 46, 109-111
Chenopodium
    album, 111
    fremontii, 111
    graveolens, 111
chess, downy, 62
chickweed, mouse-ear, 128

chicory, 77
chicory-lettuce, 77
Chimaphila umbellata, 136
chimingbells, 124
Chloris verticillata, 57
chokecherry, 44
cholla
    cane, 70
    club, 70
    walking-stick, 70
chrysanthemum, wild, 91
Chrysopsis villosa, 92
Chrysothamnus nauseosus, 46-47
Cichorium intybus, 77
Cinquefoil, 170
Cirsium
    pallidum, 83
    undulatum, 83
clammyweed, 127
cleavers, 172
Clematis
    ligusticifolia, 13
    pseudoalpina, 12
clematis, Rocky Mountain, 12
Cleome serrulata, 127
cliffbrake, 73
cliffbush, 38-39
clover
    purple prairie, 149-150
    red, 148
    white, 148
    white prairie, 149-150
cocklebur, 81
Coffeeberry family, 35, 37
columbine
    Colorado, blue, 166
    little red, 166
Commelina dianthifolia, 100
COMMELINACEAE, 100
common purslane, 107
COMPOSITAE, 42, 44-45 46-47, 74-98, 110
coneflower
    cutleaf, 88
    prairie, 87
Conium maculatum, 182
CONVOLVULACEAE, 14, 16, 48, 130
Convolvulus arvensis, 15-16, 130
Conyza canadensis, 93
Corallorhiza
    maculata, 49, 104-105
    striata, 49, 104-105
coralroot

spotted, 49, 104-105
striped, 49, 104-105
corkbark fir, 19
CORNACEAE, 23
Cornus stolonifera, 23
Corydalis aurea, 137
Coryphantha vivipara, 71
cosmos, 94
Cosmos parviflorus, 94
cota, 82
cottonwood
    narrowleaf, 25-26
    Rio Grande, 24-25
cow-parsnip, 182
coyote melon, 16, 135
Crataegus erythropoda, 25, 37
croton, 108
Croton texensis, 108
crownbeard, 92
CRUCIFERAE, 131-134
Cryptantha jamesii, 125
Cucurbita foetidissima, 16, 135
CUCURBITACEAE, 16, 135
CUPRESSACEAE, 20-21, 30
currant, wax 36, 41
Cuscuta umbellata, 14, 48
cut-leaf, yellow, 82
Cymopterus bulbosus, 181
CYPERACEAE, 66-67
Cypress family, 21, 30
Cyperus esculentus, 66
Cystopteris fragilis, 72

Dactylis glomerata, 59
daisy, 97-98
Dalea formosa, 34
dandelion
    common, 75
    mountain, 75
Danthonia intermedia, 61
Datura meteloides, 179
datura, sacred, 179
dayflower, 100
Dayflower family, 100
death camas, 104
deer's ears, 138
deervetch, 147-148
Delphinium
    occidentaie, 166
    virescens, 166
Descurainia richardsonii, 134
Dithyrea wislizenii, 132

dock, curlyleaf, 109
dodder, 14, 48
Dodecatheon pulchellum, 164
Dogbane family, 121
dogbane, spreading, 121
dogweed, 85
Dogwood family, 23
dogwood, red-osier, 23
doveweed, 108
downy chess, 62
Draba aurea, 132
dragonhead, 143
dropseed
    hairy, 63
    pine, 63
    sand, 62
Douglas-fir, 18
Dyssodia papposa, 85
dwarf juniper, 30
dwarf mistletoe, 48-49

Easter daisy, 94
Echinocereus
    fendleri, 71
    triglochidiatus, 71
    virdiflorus, 70
Echinochloa crusgalli, 59
ELAEAGNACEAE, 25
Elaeagnus angustifolia, 25
elderberry, red, 33
Elm family, 27
Elymus canadensis, 55
Ephedra family, 29-30
Ephedra viridis, 29-30
EPHEDRACEAE, 29-30
Epilobium
    angustifolium, 156-157
    ciliatum, 156-157
Epipactis gigantea, 106
EQUISETACEAE, 51
Equisetum
    arvense, 51
    hiemale, 51
    laevigatum, 51
ERICACEAE, 28-29, 50
Erigeron
    divergens, 98
    flagellaris, 97
    philadelphicus, 98
    speciosus var. macranthus, 98
    subtrinervis, 98
Eriogonum

    abertianum, 162
    alatum, 162
    cernuum, 162
    jamesii, 163
    leptophyllum, 163
    racemosum, 162
Erodium cicutarium, 139
Erysimum capitatum, 132
estafiata, 80
Eupatorium herbaceum, 79
Euphorbia serpyllifolia, 107
EUPHORBIACEAE, 107, 108
Eurotia lanata, 45
evening-primrose
    cutleaf, 158
    Hooker's, 157
    prairie, 158
    white stemless, 158
Evening-primrose family, 156-158
everlasting, pearly, 84

FAGACEAE, 24, 43-44
Fairy slipper, 105
Fallugia paradoxa, 41
false
    boneset, 79
    buffalo grass, 52
    helleborine, 102
    indigobush, 35
    pennyroyal, 144
    tarragon, 80
fendlerbush, 39
Fendlera rupicola, 39
Fern family, 72-73
fern
    bracken, 72
    brittle, 72
    lady fern, 73
fetid-marigold, 85
Figwort family, 15, 174-178
fir
    corkbark, 19
    white, 19
fireweed, 156
firewheel, 87
Flax family, 151
flax
    New Mexico yellow, 151
    Western blue, 151
fleabane
    common, 98
    daisy, 98
    Oregon, 98
    trailing, 97
Forestiera neomexicana, 38

Forget-me-not family, 124
forget-me-not, stickseed, 124
four-o'clock
    desert, 155
    showy, 155
    vining, 14
Four-o'clock family, 14, 155
four-wing saltbush, 46
Fragaria americana, 169
Franseria acanthicarpa, 81
FUMARIACEAE, 137
Fumitory family, 137

G*aillardia pulchella, 87*
Galium
    aparine, 172
    boreale, 172
    triflorum, 172
galleta, 56
gaura
    scarlet, 156
    tall, 156
Gaura
    coccinea, 156
    parvifolia, 156
gayfeather, dotted, 84
gentian
    prairie, 138
    rose, 138
Gentiana
    affinis, 138
    strictiflora, 138
Gentian family, 138
GENTIANACEAE, 138
GERANIACEAE, 139
Geranium
    caespitosum, 139
    richardsonii, 139
Geranium family, 139
Geranium
    James, 139
    Richardson's, 139
Geum triflorum, 170
giant rattlesnake plantain, 106
gilia, 161
globe mallow, 153
Glycyrrhiza lepidota, 150
goathead, 187
goatsbeard, 76
golden aster, 93
golden-eye, 86
golden-pea, big, 146
goldenrod

alpine, 90
dwarf, 90
few-flowered, 90
western, 90
goldenweed, 92
spiny, 93
Goodyera oblongifolia, 106
gooseberry, 36
Goosefoot family, 46, 109, 110, 111
goosefoot, Fremont, 111
goosegrass, 172
Gourd family, 16, 135
grama
    black, 58
    blue, 58
    hairy, 58
    side-oats, 56
GRAMINEAE, 52-64
grape, canyon, 13
Grape family, 13-14
Grass family, 52-64
grass
    barnyard, 59
    false buffalo, 52
    Indian, 63
    needle-and-thread, 60
    orchard, 59
greenthread, 85
Grindelia aphanactis, 82
groundcherry, 180
groundsel
    Bigelow, 83
    threadleaf, 87, 89
gumweed, 82
Gutierrezia sarothrae, 44-45, 86
GUTTIFERAE, 140

Habenaria sparsiflora, 105
hackberry, netleaf, 27
Hackelia floribunda, 124
Haplopappus
    gracilis, 93
    spinulosus, 93
harebell, 126
hawkweed, Fendler's, 76
hawthorn, 25, 37
healall, 142
Heath family, 28-29, 50, 136
Hedeoma drummondii, 144
hedgehog
    claret-cup, 71
    Fendler's, 71
    green-flowered, 70
Helianthella quinquenervis, 86
Helianthus

    annuus, 88
    petiolaris, 88
helleborine, 106
hemlock
    poison, 182
    water, 182
Heracleum lanatum, 182
heronsbill, 139
Heuchera parvifolia, 173
hiddenflower, 125
Hieracium fendleri, 76
Hilaria jamesii, 56
Holodiscus dumosus, 42
honeysuckle, bearberry, 40
Honeysuckle family, 33, 39, 40
hop, American, 12
hoptree, narrowleaf, 22-23, 32
horehound, 143-144
horsebrush, 47
horsemint, 142
horsetail
    meadow, 51
    smooth, 51
Horsetail family, 51
horseweed, 93
Humulus americanus, 12
HYDROPHYLLACEAE, 141
Hydrophyllum fendleri, 141
Hymenopappus filifolius, 82
Hymenoxys
    acaulis, 92
    argentea, 92
    richardsonii, 92
Hypericum formosum, 140

Indian
    grass, 63
    hemp, 121
    ricegrass, 58-59
    wheat, 160
indigobush
    false, 35
    feather, 34
Ipomoea coccinea, 16, 130
Ipomopsis
    aggregata, 161
    longiflora, 161
IRIDACEAE, 99-100

Iris family, 99-100
Iris missouriensis, 99
iris, wild, 99
Iva xanthifolia, 110
Ivy, poison, 33

Jacob's ladder, 161
Jamesia americana, 38-39
JUNCACEAE, 65-66
Juncus
    bufonius, 66
    drummondii, 66
    interior, 66
junegrass, 62
juniper
    alligator, 20-21
    dwarf, 30
    mistletoe, 48
    one-seed, 21
    Rocky mountain, 21
Juniper family, 20-21, 30
Juniperus
    communis, 30
    deppeana, 21
    monosperma, 21
    scopulorum, 21

Kinnikinnik, 28-29
kittentails, 175
Kochia scoparia, 110
Koeleria cristata, 62
Kuhnia chlorolepis, 79

LABIATAE, 142-144
Lactuca pulchella, 77
lady fern, 73
lamb's quarters, 111
Lappula redowskii, 125
larkspur
    tall purple, 166
    white, 166
Lathyrus arizonicus, 149
LEGUMINOSAE, 30-31, 34-35, 145-150
lemita, 32
lemonade berry, 32
Lepidium medium, 133
Lesquerella intermedia, 134
lettuce
    chicory, 77
    wire, 77
Leucelene ericoides, 95

Liatris punctata, 83-84
licorice, wild, 150
Ligusticum porteri, 18
LILIACEAE, 30, 101-104
Lilium umbellatum, 102
lily
    mariposa, 101
    sego, 101
Lily family, 30, 101-104
limber pine, 17
LINACEAE, 151
Linum
    lewisii, 151
    neomexicana, 151
Lithospermum
    incisum, 125
    multiflorum, 125
LOASACEAE, 152
Lobelia cardinalis, 126
lobelia, scarlet, 126
locust, New Mexico, 30-31
locoweed, 150
Lolium perenne, 56
Lonicera involucrata, 40
LORANTHACEAE, 48-49
Lotus wrightii, 147-148
lousewort, 177
lovage, Porter's, 183
lupine
    king's, 145
    tall, 146
Lupinus
    caudatus, 145-146
    kingii, 145-146
Lycium pallidum, 35-36
Lycurus phleoides, 54
lyre leaf, 91

Machaeranthera
    bigelovii, 96
    tanacetifolia, 96
Madder family, 172
Malaxis soulei, 105
mallow
    globe, 153
    white prairie, 153
Mallow family, 153-154
Malva
    neglecta, 154
    parviflora, 154
MALVACEAE, 153-154
maple
    boxelder, 22
    Rocky mountain, 23
Maple family, 22-23

marigold, fetid, 85
mariposa, lily, 101
Marrubium vulgare, 144
marsh-elder, 110
Maurandya antirrhiniflora, 15
meadowrue, Fendler, 168
Medicago sativa, 147
Melampodium leucanthum, 95
melilotus
    albus, 147
    officinalis, 146
Mentha arvensis, 143
Mentzelia
    albicaulis, 152
    pumila, 152
Mertensia
    franciscana, 124
    lanceolata, 124
milkvetch
    Missouri, 150
    stinking, 150
milkweed
    broad-leaved, 122
    poison, 123
    showy, 122
Milkweed family, 122
Mimulus
    glabratus, 177
    guttatus, 177
mint, 143
Mint family, 142-144
Mirabilis
    multiflora, 155
    oxybaphoides, 14
mistletoe
    dwarf, 48-49
    juniper, 48
Mistletoe family, 48-49
mockorange, 40
Moldavica parviflora, 143
Monarda
    menthaefolia, 142
    pectinata, 143
monkeyflower, 177
monkshood, 165
Monotropa latisquama, 50
monument plant, 138
MORACEAE, 12
Mormon tea, 29-30
Morning-glory family, 14, 16, 48, 130

mountain
    dandelion, 75
    lover, 29
    mahogany, 38, 42-43
    muhly, 63
    parsley, 181
Muhlenbergia montana, 63
muhly, mountain, 63
Mulberry family, 12
mullein, 174
Munroa squarrosa, 52
mustard
    tansy, 134
    tumble, 134
Mustard family, 131-134

Narrowleaf
    hoptree, 22-23, 32
    cottonwood, 26
    yucca, 101-102
Navajo tea, 82
needle-and-thread grass, 60
needlegrass, 60
Nettle family, 108
nettle, stinging, 108
New Mexico
    locust, 30-31
    olive, 38
    porcupine grass, 60
    yellow flax, 151
nightshade
    black, 180
    silverleaf, 179
Nightshade family, 36, 179-180
ninebark, 41
nodding
    onion, 103
    wood sunflower, 86
nutsedge, yellow, 66
NYCTAGINACEAE, 14, 155

Oak
    Gambel, 24, 43-44
    gray, 43-44
    wavyleaf, 43-44
oatgrass, timber, 61
ocean-spray, 42
Oenothera
    albicaulis, 158
    caespitosa, 158
    coronopifolia, 158
    hookeri, 157
old-man's whiskers, 170
OLEACEAE, 38

Oleaster family, 25
Olive family, 38
olive
    desert, 38
    New Mexico, 38
    Russian, 25
ONAGRACEAE, 156-158
onion
    Geyer, 103
    nodding, 103
Opuntia
    clavata, 70
    erinacea, 69
    imbricata, 70
    macrorhiza, 68
    phaeacantha, 69
    polyacantha, 69
    rhodantha, 69
orchard grass, 59
ORCHIDACEAE, 49, 104-106
orchid
    bog, 105
    fringed, 105
    coralroot, 49, 104-105
orchid family, 49, 104-106
OROBANCHACEAE, 50
Orobanche fasciculata, 50
Orthocarpus
    luteus, 177
    purpureo-albus, 175
Oryzopsis hymenoides, 58-59
osha, 183
Osmorhiza obtusa, 181-182
owl-clover
    purple-white, 175
    yellow, 177
OXALIDACEAE, 159
Oxalis family, 159
Oxalis violacea, 159
Oxybaphus linearis, 155

Pachystima myrsinites, 29
paintbrush, 178
    foothills, 178
    scarlet, 178
    Wyoming, 178
    yellow, 178
pale trumpet, 161
Panicum capillare, 59
paperflower, wooly, 91
Parthenocissus inserta, 14
pasque flower, 167
Pea family, 30-31, 34, 35, 145-150
pearly-everlasting, 84

peavine, Arizona, 149
Pectis angustifolia, 85
Pedicularis grayi, 177
Pellaea spp., 73
pennycress, 132
pennyroyal false, 144
penstemon
    beardtongue, 176
    James, 176
    variegated, 176
    Whipple's, 176
Penstemon
    barbatus var. *torreyi, 176*
    jamesii, 176
    secundiflorus, 176
    virgatus, 176
    whippleanus, 176
peppergrass, 133
Pericome caudata, 78
perky sue, 92
Petalostemum
    candidum, 149-150
    purpureum, 149-150
Phacelia
    corrugata, 141
    heterophylla, 141
Philadelphus microphyllus, 40
Phleum pratensis, 54
Phlox family, 161
Phoradendron juniperinum, 48
Physalis foetens var. *neomexicana, 180*
Physocarpus monogynus, 41
Picea
    engelmannii, 20
    pungens, 19-20
pigweed, 108
PINACEAE, 17-20
pincushion cactus, 71
pine
    limber, 17
    pinyon, 17
    ponderosa, 18
Pine family, 17-20
pinedrops, 50
pinesap, 50
Pink family, 128-129
Pinus
    edulis, 17
    flexilis, 17
    ponderosa, 18
pipsissewa, 136
PLANTAGINACEAE, 160

Plantago
    major, 160
    psyllium, 160
    purshii, 160
Plantain family, 160
plantain
    flaxseed, 160
    giant rattlesnake, 106
    rippleseed, 160
plum, wild, 37
plume, Apache, 41
poison ivy, 33
Polanisia trachysperma, 127
POLEMONIACEAE, 161
Polemonium foliosissimum, 161
POLYGONACEAE, 15, 109, 162-163
Polygonum
    aviculare, 162
    convolvulus, 15
POLYPODIACEAE, 72-73
Polypogon monspeliensis, 55
ponderosa pine, 18
ponymint, 143
Populus
    angustifolia, 25-26
    fremontii, 24-25
    tremuloides, 24
porcupine grass, New Mexico, 60
Porter's lovage, 183
PORTULACACEAE, 107
Portulaca oleracea, 107
Potentilla
    anserina, 170-171
    fruticosa, 34
    hippiana, 170-171
    norvejica, 170-171
    pulcherrima, 170-171
potentilla
    beauty, 171
    Norway, 171
    shrubby, 34
    silvery, 171
poverty three-awn, 60
prairie
    coneflower, 87
    gentian, 138
prairie clover
    purple, 149-150
    white, 149-150
prickly pear

    cliff, 69
    hedgehog, 69
    purple fruit, 69
    starvation, 69
    tuberous, 68
Primrose family, 164
PRIMULACEAE, 164
Prunella vulgaris, 142
Prunus
    americana, 37
    virginiana var. *melanocarpa, 44*
Pseudocymopterus montanus, 181
Pseudotsuga menziesii, 18
Psilostrophe tagetina, 91
Ptelea trifoliata , 22-23, 32
Pteridium aquilinum, 72
Pterospora andromedea, 50
puccoon, 125
Pulsatilla ludoviciana, 167
puncture vine, 187
purslane, common, 107
Purslane family, 107
pussytoes, 84
pyrola, 136
Pyrola chlorantha, 136

Quassia family, 22
Quercus
    gambelii, 23-24, 43-44
    grisea, 43-44
    undulata, 43- 44

Rabbitbrush, rubber, 47
rabbitfoot grass, 55
ragweed, 81
Ramischia secunda, 136
RANUNCULACEAE, 12-13, 34, 165-168
Ranunculus
    aquatils, 166-167
    cardiophyllus, 166-167
    inamoenus, 166-167
    macounii, 166-167
raspberry, wild, 31
Ratibida
    columnifera, 87
    tagetes, 87
RHAMNACEAE, 35, 37
Rhus
    radicans, 33
    trilobata, 32
Ribes

    cereum, 36, 41
    inerme, 36
ricegrass, Indian, 59
Rio Grande cottonwood, 24-25
Robinia neomexicana, 30-31
rockcress, Fendler, 133
rock-jasmine, 164
Rocky Mountain
    bee-plant, 127
    clematis, 12
    lily, 102
    maple, 23
    sage, 144
Rorippa nasturtium-aquaticum, 131
Rosa woodsii
    var.*fendleri, 31-32*
    var *hypoleuca, 31-32*
ROSACEAE, 25, 31-32, 34, 37, 38, 40-44, 169-171
rose
    Fendler's, 31-32
    gentian, 138
    paleleaf, 31-32
    wild, 31
Rose family, 25, 31-32, 34, 37, 38, 40-44, 169-171
RUBIACEAE, 172
Rubus
    strigosus var. *arizonicus, 31*
    parviflorus, 40
Rudbeckia
    hirta, 88
    laciniata, 88
Rue family, 22-23, 32
Rumex
    acetosella, 109
    crispus, 109
Rush family, 65-66
Rush
    drummond, 66
    inland, 66
    toad, 66
Russian
    olive, 25
    thistle, 109
    wheatgrass, 54
RUTACEAE, 22-23, 32
ryegrass, perennial, 56

Sagebrush
    big, 42, 80
    Bigelow, 80
    ragweed, 80
    sand, 47
Saint Johnswort, 140
Saint Johnswort family, 140
SALICACEAE, 24-26, 45-46
Salix
    bebbiana, 45-46
    exigua, 45-46
    irrorata, 45-46
salsify, yellow, 76
Salsola kali, 109
saltbush, four-wing, 46
salvia, 144
Salvia reflexa, 144
salt-cedar, 20
Sambucus microbotrys, 33
sand
    aster, 95
    sagebrush, 47
sandbur, field, 53
sandwort, Fendler's, 129
Saxifraga
    bronchialis, 173
    rhomboidea, 173
SAXIFRAGACEAE, 36, 38-41, 173
Saxifrage family, 36, 38-41, 173
saxifrage, spotted, 173
scarlet bugler, 176
scorpionweed, 141
scouring rush, 51
SCROPHULARIACEAE, 15, 174-178
Sedge family, 66-67
sedge, 66
    field, 67
    meadow, 67
sego lily, 101
selfheal, 142
Senecio
    bigelovi, 83
    douglasii, 87
    eremophilus var. *macdougalii*, 89
    fendleri, 89
    multicapitatus, 89
    multiflorus, 90
    neomexicanus, 90
serviceberry, 43
Setaria viridis, 54-55
sheep sorrel, 109
shepherd's purse, 133
showy milkweed, 122
sidebells, 136
Sidalcea candida, 153

Silene scouleri, 128
silverweed, 171
SIMAROUBACEAE, 22
Sisymbrium altissimum, 134
Sisyrinchium
    demissum, 100
    montanum, 100
Sitanion hystrix, 55
skunk cabbage, 102
skunkbush, 32
skyrocket, 161
slipper, fairy, 105
Smilacina
    racemosa, 103
    stellata, 103
snakeweed, 44-45, 86
snapdragon vine, little, 15
snowberry, 39
SOLANACEAE, 36, 179-180
Solanum
    elaeagnifolium, 179
    nigrum, 180
    rostratum, 180
Solidago
    multiradiata, 90
    occidentalis, 90
    sparsiflora, 90
    spathulata, 90
Solomon's
    plume, 103
    seal, false, 103
Sonchus asper, 76
Sorghastrurn nutans, 63
sorrel, sheep, 109
sow-thistle, spiny-leaved, 76
spectacle pod, 132
speedwell, American, 175
Sphaeralcea
    coccinea, 153
    incana, 153
Sporobolus cryptandrus, 62
spruce
    Colorado blue, 19-20
    Engelmann, 20
Spurge family, 107-108
spurge, thymeleaf, 107
squaw lettuce, 141
squawbush, 32
squirreltail, bottlebrush, 55
Staff-tree family, 29
star
    white-stemmed blazing, 152
    glory, 16, 130
    flower, 103
starwort, 129
Stellaria jamesiana, 129

Stephanomeria
    pauciflora, 77
    tenuifolia, 77
stickleaf, 152
stickseed, 125
stinging nettle, 108
stinkweed, 127
Stipa
    comata, 60
    neomexicana, 60
strawberry, wild, 169
Sumac family, 32-33
summer-cypress, 110
sundrops, 157
sunflower, 88
    annual, 88
    prairie, 88
Sunflower family, 42, 44, 46-47, 74-98, 110
sweet cicely, bluntseed, 182
sweet clover
    white, 147
    yellow, 146
Swertia radiata, 138
Symphoricarpos oreophilus, 39

TAMARICACEAE, 20
tamarisk, 20
Tamarisk family, 20
Tamarix pentandra, 20
tansy mustard, 134
taperleaf, 78
Taraxacum officinale, 75
tarragon, false, 80
tea
    Mormon, 29-30
    Navajo, 82
Tetradymia canescens, 47
Texas timothy, 54
Thalictrum fendleri, 168
Thelesperma
    megapotamicum, 82
    trifidum, 85
Thelypodium wrightii, 133
thelypody, 133
Thermopsis pinetorum, 146
Thlaspi alpestre, 132
thoroughwort, 79
three-awn

Arizona, 60
poverty, 60
red, 60
thymeleaf spurge, 107
timothy, 54
　Texas, 54
tobacco root, 184
tomatillo, 36
Townsendia
　exscapa, 94
　incana, 95
Townsend's aster, 95
Tragopogon
　dubius, 76
　pratensis, 76
tree-of-heaven, 22
Tribulus terrestris, 187
Trifolium
　pratensis, 148
　repens, 148
tumble mustard, 134
turkeyfoot, 57
Typha latifolia, 65
TYPHACEAE, 65

ULMACEAE, 27
UMBELLIFERAE, 181-183
Urtica gracilis, 108
URTICACEAE, 108

Vaccinium myrtillus, 29
Valerian capitata, 184
Valerian family, 184
VALERIANACEAE, 184
Veratrum californicum, 102
Verbascum thapsus, 174
Verbena
　bracteata, 185
　macdougalii, 185
　wrightii, 185
VERBENACEAE, 185
Verbesina encelioides, 92
Veronica americana, 175
vervain, 185
Vervain family, 185
vetch
　American, 149
　deer, 147-148
Vicia americana, 149

Viguiera multiflora, 86
vine, puncture, 187
vining four-o'clock, 14
Viola
　adunca, 186
　canadensis, 186
　nephrophila, 186
　pedatifida, 186
VIOLACEAE, 186
violet
　northern bog, 186
　Canada, 186
　larkspur, 186
　western dog, 186
　woodsorrel, 159
Violet family, 186
virgin's bower, western, 13
Virginia creeper, 14
VITACEAE, 13-14
Vitis arizonica, 13

Wafer-parsnip, 181
wallflower, western, 132
water-birch, 27
water-crowfoot, 167
watercress, 131
Waterleaf family, 141
wheatgrass
　desert, 54, 57
　Russian, 54
　slender, 57
　western, 57
white fir, 19
whitlow grass, 132
whortleberry, 29
wild
　buckwheat, 162
　candytuft, 132
　chrysanthemum, 91
　onion, 103
　plum, 37
　strawberry, 169
wildrye, Canada, 55
Willow family, 24-26, 45-46
willow
　Bebb, 45-46
　bluestem, 45-46
　coyote, 45-46
willowweed, 156-157
windmill grass, 57
winterfat, 45

wintergreen, 136
wire-lettuce, 77
witchgrass, 59
wolfberry, pale, 35-36
wolftail, 54
wood lily, 102
woodsorrel, violet, 159
wood-sunflower, nodding, 86
woolly
　Indian-wheat, l60
　paperflower, 91
wormwood, 79-80

Xanthium strumarium, 81

Yarrow, 93
yucca, 30
　banana, 101-102
　narrowleaf, 101-102
Yucca
　angustissima, 101-102
　baccata, 101-102

Zygadenus elegans, 104
ZYGOPHYLLACEAE, 187

208

NOTES

NOTES